THE USUFRUCTUARY ETHOS

The Usufructuary Ethos

POWER, POLITICS, AND ENVIRONMENT IN THE LONG EIGHTEENTH CENTURY

Erin Drew

UNIVERSITY OF VIRGINIA PRESS

CHARLOTTESVILLE AND LONDON

University of Virginia Press
© 2021 by the Rector and Visitors of the University of Virginia
All rights reserved
Printed in the United States of America on acid-free paper

First published 2021

1 3 5 7 9 8 6 4 2

Library of Congress Cataloging-in-Publication Data

Names: Drew, Erin E., author.
Title: The usufructuary ethos : power, politics, and environment in the long
eighteenth century / Erin Drew.
Description: Charlottesville : University of Virginia Press, 2021. | Includes
bibliographical references and index.
Identifiers: LCCN 2020051266 (print) | LCCN 2020051267 (ebook) |
ISBN 9780813945798 (hardcover ; acid-free paper) | ISBN 9780813945804
(paperback ; acid-free paper) | ISBN 9780813945811 (ebook)
Subjects: LCSH: Environmentalism in literature. | Literature and
society—Great Britain—History—18th century. | Literature and society—
Great Britain—History—17th century. | Nature—Effect of human
beings on—Great Britain. | Ecocriticism—Great Britain.
Classification: LCC PR448.E58 D74 2021 (print) | LCC PR448.E58 (ebook) |
DDC 820.9/36—dc23
LC record available at https://lccn.loc.gov/2020051266
LC ebook record available at https://lccn.loc.gov/2020051267

Cover art: "Plate 10, Outlines of Figures, Landscapes and Cattle . . . for the Use of
Learners," Thomas Rowlandson, hand-colored etching, 1790 (The Metropolitan
Museum of Art, NY, www.metmuseum.org, The Elisha Whittelsey Collection,
The Elisha Whittelsey Fund, 1959); background, Sonate/iStock

CONTENTS

THE USUFRUCTUARY ETHOS

INTRODUCTION

"Usufruct" and the Eighteenth Century in Environmental Criticism

IN HIS 2016 book *Dangerous Years: Climate Change, the Long Emergency, and the Way Forward*, the political scientist and environmental activist David Orr proposes looking to a concept called "usufruct" and a series of eighteenth-century figures to understand the intergenerational responsibility demanded by the twenty-first century's environmental crisis. Deriving from Roman property law, "usufruct" refers to the "right of temporary possession, use, or enjoyment of the advantages of property belonging to another, so far as may be had without causing damage or prejudice."[1] Orr quotes Thomas Jefferson's famous claim from a 1789 letter to James Madison that "the earth belongs in usufruct to the living," then connects it to John Locke's assertion in the *Second Treatise on Civil Government* that just use of natural resources must leave "enough and as good" behind for others.[2] Orr muses:

> Are there reasons persuasive and powerful enough to override the perceived self-interest of entire generations that would compel them to leave "as much and as good" for subsequent generations? A great deal depends on how we answer that question, so let me offer two and a half possible answers. The first is drawn from the great conservative Edmund Burke, who wrote, "People will not look forward to posterity, who never look backward to their ancestors." In his view, a chain of obligation connects past, present, and future generations and when honored can help overcome the "selfish temper and confined views" of the present—that is, it can work in our self-interest.[3]

Within the eighteenth century's usufructuary ideas, Orr suggests, lies a lesson about the responsibilities we have toward each other, the world, and the future that is congenial to our contemporary environmental concerns. Usufruct could provide a paradigm that balances use with sustainability, personal interest with public welfare, the demands of the present with the needs of the future.

What Orr did not realize was that an environmental ethos along the lines he describes long predated Jefferson and Burke. *The Usufructuary Ethos* recovers usufruct's legacy as an influential way of understanding the moral relationship between humans and their environments in late seventeenth- and early eighteenth-century England.[4] During this period, this book will demonstrate, "usufruct" appears as a common point of reference and comparison across philosophical, devotional, legal, and literary discussions of the ethical parameters of possession, use, and power. The way that usufruct positions possession as temporary and limited allowed writers to articulate the interlocking "chains of obligation" that made up their world: chains that linked not only past, present, and future but humans, nonhumans, and God, as well as the social, political, and natural worlds. Burke did not have the natural world in mind when he spoke of posterity in the sentence quoted by Orr, but the fact that his words apply so neatly to environmental ideas is less a coincidence than an echo of the legacy of the usufructuary socio-environmental ethos of earlier decades. Given that the usufructuary ethos was strongly associated with politically conservative writers such as John Evelyn, Anne Finch, and Alexander Pope—writers dedicated to the political and moral centrality of the landed gentry and the land itself—Burke's late-century invocation of usufructuary ideas to defend conservative political ideologies indicates his debt to the usufructuary ethos. *The Usufructuary Ethos* recenters the titular concept in late seventeenth- and eighteenth-century culture, and makes a case for it as an environmental ethic unique to that period in England.

I use the word "ethos" to describe this thread of early modern thought not only to illustrate the way that "usufruct" provided a set of guiding beliefs for a significant group of English writers but also to emphasize its ethical impetus and ramifications. The premise that humans had only usufruct of the earth, rather than full control or ownership of it, provided a framework for determining the best uses of the nonhuman world, not in terms of what sort of use was most productive, but in terms of what sort of use would best fulfill the user's responsibilities to others, both human and nonhuman, in the present and future. Usufructuary discourse is therefore distinct from the better-known contemporaneous discourse of "improvement," which sought to maximize the productive potential of the natural world through human innovation. Improvement was, of course, motivated in some cases by altruistic desires to eliminate dearth and improve life; but whatever the motivations of the individual practitioner, it was concerned with identifying the best ways for landowners to intervene in nature to "improve" its produce. In contrast, the usufructuary ethos places the human owner of land in the middle of a hierarchy, responsible to God for the natural gifts he has been granted (and which God retains ownership of), and responsible to other beings and future generations for passing along intact what they will need to survive. The usufructuary ethos sees the world as an interspecies, intergenerational network of obligation and dependence in which each organism carries moral significance.

Alongside "usufruct," a network of other, associated legal metaphors such as *steward, tenant,* and *landlord* took on specific significance within the discourse of the usufructuary ethos. These terms share a few crucial commonalities. First, they describe hierarchical relationships organized around property. The usufructuary, the trustee, and the steward all have in common the fact that they wield power over property that does not belong to them. They also have moral and fiduciary obligations to that property's current and future owners to maintain and care for it. These figures are, therefore, doubly medial: in the middle of a social and legal chain between proprietor and dependent, and in the middle of a temporal chain between past and future. Second, usufructuary, steward, tenant, and landlord are all terms for legal, social, and political relationships of power that were applied to relationships among nonhuman beings and God as well as to relationships among humans. What this reveals, I argue, is that the usufructuary ethos represented a moral framework that applied to all aspects of existence, "natural" and "social" alike. Writers used analogies borrowed from nature as evidence for moral obligations in the social world, and vice versa; this was evidence, not of an attempt to impose human mastery over nature, but rather of the conviction that moral obligations to both nature and other humans derived from the same source.

Thus, this book yields an important insight for environmental studies: that the ways a culture defines justice and the ethics of power and use in the political and social realms are fundamentally connected to the ways it understands environmental ethics. *The Usufructuary Ethos* is both a study of an important but largely forgotten historical environmentalism unique to seventeenth- and eighteenth-century England and a case study in the implications for environmental thought of one important early modern theory of legitimate and illegitimate uses of power, in which legitimate power is defined by accountability and adherence to moral duty. The overlapping vocabularies of philosophical, legal, and religious discourse point to the cultural pervasiveness of the usufructuary distinction between just and legitimate uses of land and power and unjust and illegitimate ones. Borrowing language from the concept of usufruct to distinguish God's absolute dominion over creation from mankind's usufructuary dominion offered a way for late seventeenth- and early eighteenth-century writers to persuade readers that humans were morally accountable for the care and preservation of creation.

As is so often the case, however, it only became necessary to explicitly articulate the usufructuary ethos in the context of competing ideas about the relationship between humans and nature. One of them was the oft-cited narrative that "man's authority over the natural world was . . . virtually unlimited," and that English theological and moral writing was so heavily focused on humankind's unique dominion that, as Keith Thomas once wrote, its readers "could be forgiven for inferring that their main purpose was to define the special status of man and to justify his rule over other creatures."[5] Such attitudes to nature were long accepted as representative of the period by ecocritics; *The Usufructuary Ethos*, among other things, offers a corrective to that oversimplified view. The usufructuary ethos was a powerful, widely disseminated counternarrative to such beliefs, one that insisted on stewardship and responsibility as moral obligations attached to the ontological, social, and material advantages held by human beings. The ideas of displaced ownership, mediality, and accountability at the heart of the usufructuary ethos made it possible for seventeenth- and eighteenth-century English writers to conceive of human beings as being masters and subjects at the same time, and provided an ethical framework that required human beings to act with foresight and care.

A more important factor in the usufructuary ethos's centrality in late seventeenth- and early eighteenth-century English culture, however, was the fact that it offered the writers who invoked it a bulwark rooted in older, traditional morals against the social, political, and economic changes that were rapidly transforming the world around them. Ironically, the main drivers of these changes can be traced in David Orr's two other eighteenth-century sources for the idea of usufruct: Thomas Jefferson and John Locke. The letter in which Jefferson declared that "the earth belongs in usufruct to the living" was not, in fact, about the earth at all. It was about monetary debt. Jefferson borrowed the concept of usufruct in order to set up his argument against fiscal debt on the basis of (financial) justice to future generations. In so doing, Jefferson used a rhetorical move also found in eighteenth-century poetry, borrowing the logic and, more importantly, the moral authority of the usufructuary ethos to defend economic and agricultural innovations that materially benefited him. The roots of this rhetorical move can be traced back though Locke's *Second Treatise*. Though his theory of property had its origins in the same usufructuary philosophical tradition as the usufructuary ethos, the first chapter of this book will show how Locke's theory of money and his construction of an unappropriated "America" enabled him to set aside the ethical limits imposed on the use of nature by the usufructuary ethos. Indeed, the twin specters of the financial revolution and mercantile colonial expansion shadow the texts explored throughout this book. They provided both the impetus for authors' insistent and explicit articulation of the usufructuary ethos in this particular time and place and the undertow of cultural and material transformation that would eventually de-center it.

The Usufructuary Ethos tells the story of a key moment in social and natural history during which changes accelerated that would lead to the Anthropocene: economic shifts into a new form of world capitalism, into imperialism, into liberalism; changes in formations of knowledge that would split "nature" from "society" in unprecedented ways; and in England itself, the final stages of enclosure and the arrival of the Industrial Revolution. This is a story about an old idea, usufruct, that took on, for a while, a central role in articulating the ethical relationships among living beings in a world that seemed to be slowly disintegrating those relationships. This is a story that has been mostly forgotten, I believe, because its logic and structure derive from sources twenty-first-century scholars have not been in the habit of looking to for environmental history: popular theology, moral philosophy, law. This is a story that was repeated to the English population over and over again in sermons and devotional tracts, forms dedicated to exploration of moral and ethical duties to God, society, *and* nature.

Finally, it is a story that unfolds in its most thoughtful, complex, and revealing guise in poetry. Following Richard Feingold, I am interested in the "intersection of art and actuality," and in resurrecting the places where "forms of social and political [and environmental] understanding" and "attitudinal and expressive habits" cross the generic boundaries between philosophy, theology, law, and poetry.[6] The portions of this book that consider prose like the devotional works of Richard Allestree or John Evelyn's silviculture manual *Silva* are not merely background for the poetry, but examples of forms of ethical and environmental thought shared across generic lines. Still, poetry is unique in its capacity to contain the ideological tension of simultaneously held yet incompatible beliefs, and thereby to lay bare in its full complexity the experience of living through profound cultural and material transformations while clinging to continuity. By tracing the usufructuary ethos's rise and fall through poetry, this book aims to create a better understanding not only of the environmental thought of the eighteenth century itself but of the ways a culture in the midst of environmental transformation attempts imaginatively to reckon with itself.

LANDLORDS, POLITICS, AND ENVIRONMENTAL ETHICS OF EIGHTEENTH-CENTURY ENGLAND

The fact that the ontological and moral structures of the usufructuary ethos hold for relationships among humans and nonhumans (what we might call "nature") as well as those among humans ("politics" or "society") has important implications for the interconnections of political and environmental thought in the long eighteenth century. The analogy between humans as lords over nature and landlords as lords over other people has most commonly been explained as a function of the attempt to preserve and "naturalize" the hegemonic power of the gentleman. In John Barrell's influential account, descriptive poetry of the eighteenth century revealed a "desire to impose an order on a landscape, by laying structure on it" in order to assert authority over nature itself.[7] Part of the impetus for that desire, he later elaborated, came from the destabilization of the authority of the landed gentleman as a result of the political and economic shifts of the Walpole years.[8] Furthermore, "as enclosure, improvement, and more commercial estate management replaced a more traditional paternalist authority," Tim Fulford later wrote, there was a corresponding loss of the "commanding [prospect] view" as "either purely aesthetic or an emblem of" gentlemanly authority.[9]

Barrell's and Fulford's critiques, while powerful and productive, exemplify a critical practice that eventually produced two roadblocks for eighteenth-century environmental literary criticism. First, they read natural imagery in poetry and literature as primarily emblematic of social or political positions rather than as descriptions of actual natural scenes or explorations of the relationships among human and nonhuman worlds. And second, they present eighteenth-century ideas of hierarchy in their most straightforward, pejorative form, as expressions of mastery and the desire to keep and exert control. The "hermeneutics of suspicion" led a generation of critics to take English poets' and devotional writers' depictions of nature as entirely discursive, social constructions projected onto the natural world for their own ideological ends. When eighteenth-century writers "say they are 'finding out' moral lessons *in* nature," Courtney Weiss Smith has recently written, "scholars assume that they are really projecting meanings *onto* nature, 'misrecognizing' their own desires, or cunningly conscripting nature's authority to serve interested ends."[10] The a priori critical assumption that nature and society are separate domains results, in part, in a scholarly tendency to consider any early modern writer's claim to have located moral guidance in nature inherently invalid: confused, or cynical, or both. Smith goes on to note that the "'modern' reading practices" that have "carve[d] up the world into mutually exclusive categories of social and natural" have been "especially influential in readings of eighteenth-century religion and poetry,"[11] which has further contributed to scholars ignoring or eschewing the environmental significance of devotional writing.

The Usufructuary Ethos proposes a different way of understanding the hierarchism of eighteenth-century environmental thought. The usufructuary ethos is an ethic of use and of power, which, while it upholds the hegemonic hierarchies of early modern England, responds to the inequalities inherent in those hierarchies by articulating moral rules over use and power in order to prevent abuse and harm. On the one hand, the usufructuary ethos is built on the assumption that hierarchy is both necessary and right in society and in nature. On the other hand, it was grounded in an ethic of care that considered nonhuman beings—animals, plants, soil, even, occasionally, the sun—members of the moral universe to and for whom human beings were accountable for the exact same reasons that humans are accountable to and for other humans.[12]

The cultural, social, political, and environmental importance of the usufructuary ethos coalesced in the late seventeenth- and early eighteenth-century in the figure of the landlord: he who received the greatest gifts from God, and with them the greatest responsibilities.

The hierarchical ontology of the usufructuary ethos, this book argues, thus entails a particular sort of environmental ethic, one that is tied discursively to the political figure of the landlord. Landlordship carried with it specific social and environmental obligations, particularly to future generations ("posterity"), and to the other living beings who depended on the landlord and his lands (his "public"). The usufructuary ethos takes for granted that human beings had a greater degree of power and property than other beings in the hierarchy of being. It articulates an ethics of possession and power meant to specify under what terms and to what ends a usufructuary landlord may use his property and exercise his power, and to whom he is accountable for those uses. This ties into a broad set of socio-environmental concerns that came to a head during the mid-eighteenth century, loosely organized under the heading the "use of riches," which applied to the political, social, and environmental repercussions of misuse and abuse.

Anxieties about those repercussions intensified as the economic and material changes wrought by the financial revolution, "improvement," enclosure, and colonialism accelerated through the eighteenth century. In response, writers worked, in varying ways, to try either to defend usufructuary values against encroaching change or to reconcile the divergent modes of existence with one another. In that sense, *The Usufructuary Ethos* provides another side of the story that Barrell and Fulford told about the political, natural, and literary repercussions of these changes in English life. Part of what drove writers like John Evelyn, Anne Finch, Alexander Pope, and John Dyer to reassert the values of the usufructuary landlord in their poems was the conviction that the position of subordinate, accountable, temporary power granted to the usufructuary guaranteed the continued stability and well-being of both the social and the natural worlds. Indeed, Barrell observes that the "idea of social coherence" being achieved through personal wealth did "not become especially respectable . . . until the mid-century, for it required that self-interest, and not simply self-preservation, be now approved as necessary to the well-being of a society whose survival had earlier seemed dependent on its success in subduing self-interest to a sense of the interest of the whole. . . . [Self-interest] seems to become [respectable] only by being reattached to the virtues" already established in English society.[13]

Both Richard Feingold and Suvir Kaul have elaborated on this point, demonstrating the ways that eighteenth-century English poetry worked to marry new economic-colonial activities with older poetic forms, as well as how such poems reveal the dissonance and ambivalence at the heart of the emerging English empire.[14] *The Usufructuary Ethos* brings these insights to bear on the eighteenth-century environment, probing the ways that the economic, political, and poetic tensions revealed by these scholars were bound up with the environmental transformations, both conceptual and material, that took place at the same time. In doing so, I am connecting Feingold's and Kaul's literary arguments to more recent work by environmental historians to reconnect political and environmental histories. Most relevant among these recent histories is Fredrik Albritton Jonsson's *Enlightenment Frontiers*, which argues for the inextricability of natural and political economy in mid to late eighteenth-century Scotland. Jonsson observes that a desire among the "landed interest [to] reconstitute itself as a protector of the social order in an age of commercialization and agrarian transformation" led to a "self-conscious anachronism" in contrast to the classical liberalism of thinkers like Hume and Smith, and laid the conditions not only for political activity but for a whole way of engaging with nature.[15] My book points to a prehistory for Jonsson's, tracing some of the origins of this dynamic to early eighteenth-century English writers who reacted against emerging forces of capitalism and liberalism by reaching back to more traditional socioenvironmental ethics such as usufruct.

Recovering the usufructuary ethos enables us to see that, for many thinkers in the late seventeenth and eighteenth centuries, the superior position accorded to particular groups of human beings not only justified their possession of power but also restricted and bounded it. Their hierarchically superior position immiscibly entangled them with the rest of creation both through their commonalities as created beings *and* through their obligations to preserve the well-being of other beings. The responsibility that human beings' usufructuary power carries with it has moral and environmental implications that the previous focus on its complicity in inequality (understandably) occluded. Those implications are crucial to understanding early moderns' relationship to nature, as well as how and why that relationship changed over the course of the eighteenth century. In other words, though inescapably hierarchical in nature, the usufructuary ethos enforces an ethical relationship among human and nonhuman beings that, ideally if not always practically, implicates all members of the hierarchy in the well-being of the others. Though it differs from contemporary environmental ethics in fundamental ways, and though it will prove to be deeply flawed, it was nonetheless a real and widely held way of understanding the ethical relationship between human beings and their environment.

Power, authority, rights, ownership—all political issues—impinge upon the ways humans conceive of their relationships to the nonhuman world in which they are embedded, and the ideologies of natural and political order have, as literary critics have long recognized, been used to reinforce one another, in the seventeenth and eighteenth centuries as well as before and since. Yet the fact that nature was called upon to reinforce political and cultural ideology (and vice versa) did not preclude earnest consideration of what the human relationship with nature is or ought to be in the long eighteenth century, any more than it does in the twenty-first. The usufructuary ethos provided an ontological and ethical structure upon which writers built their understandings of human beings' fundamental relationships to every other being around them, human and nonhuman alike. Their fierce defense (or, in some cases, co-option) of these relationships indicates not only how deeply embedded the usufructuary belief was in eighteenth-century culture but also how it was falling out of sync with the world it was meant to describe.

The Usufructuary Ethos and the Histories of Eighteenth-Century Environment

For much of its history, the field of ecocriticism found it (in David Fairer's wry understatement) "difficult to gain a purchase on the eighteenth century that is anything other than negative."[16] To the first wave of ecocritics in the 1990s and early 2000s, the eighteenth century represented everything the field defined itself against: anthropocentrism, instrumentalism, the arrogant belief that humans had a right to the uninhibited exploitation of an inert material world, and, of course, the presumption of human beings' hierarchical superiority.[17] Jonathan Bate's influential *The Song of the Earth* (2000), for example, skips from the Renaissance to the Romantics, arguing that it was up to "Romanticism and its afterlife" to explore the "relationship between external environment and ecology of mind" and to grasp "poetic language as a special kind of expression which may effect an imaginative reunification of mind and nature."[18] Bate's position reflects some common assumptions of early British ecocritical thought, most particularly the strong association of Romanticism with environmentalism, on the basis of an environmental philosophy and history that blamed ecological decline on the

Enlightenment's "disenchantment" of nature, and located the possibility for "re-enchantment" in Romantic poetics.[19] Yet as many eighteenth-century critics have shown since, that thesis was based on an unfair and underinformed caricature of eighteenth-century thought and literature.[20] The oversight is "particularly surprising," John Sitter and I wrote in a 2011 essay on the case for an eighteenth-century ecocriticism, "given the venerable tradition of the study of 'Nature' in eighteenth-century studies."[21] A. O. Lovejoy's *The Great Chain of Being* (1936), Basil Willey's *The Eighteenth-Century Background: Studies on the Idea of Nature in the Thought of the Period* (1941), Marjorie Nicholson's *Mountain Gloom and Mountain Glory*, Clarence J. Glacken's *Traces on the Rhodian Shore* (1967), Keith Thomas's *Man and the Natural World* (1985): these studies detailed the rich, multifarious, complex relationships between humans and their environments in the early modern world long before ecocriticism or even (in some cases) environmentalism emerged.

Nevertheless, the first eighteenth-century scholars to venture into environmental criticism found themselves in a tough spot. On the one hand, they had to refute the consensus in the ecocritical community that "eighteenth-century" was synonymous with "anti-ecological." On the other, they had to explain to the eighteenth-century community why influential theories such as Raymond Williams's and John Barrell's, which read landscape as an emblem of politics, did not capture the whole truth about eighteenth-century literature's engagement with nature. With early ecocritical theory eschewing politics almost entirely and conflating any discussion of the use of nature with exploitation,[22] eighteenth-century scholars sought to restore to prominence expressions of delight in and respect for nature in eighteenth-century literature, while also acknowledging that "contradictory" perspectives like anthropocentrism or instrumentalism appear just as frequently, often in the very same texts.[23] This "double gesture of both deference and mastery before nature" was the "characteristic feature of eighteenth-century nature writing," Christopher Hitt wrote in 2004, in one of the earliest essays on eighteenth-century ecocriticism.[24] Yet to argue that eighteenth-century depictions of nature are characteristically contradictory is to continue to judge the period's environmental thought according to first-wave ecocriticism's anachronistic and distorting dualism. That divide is a legacy of an earlier phase of environmentalism, when conservationists could, at least some of the time, imagine that their concerns transcended and were separate from the human realm of politics, when they could imagine that the human was separable from Nature, and that Nature must be insulated from the human to be itself.

The last decade, however, has seen a flowering of scholarly interest in examining the numerous and complicated ways eighteenth-century writers grappled both directly and imaginatively with their environments. David Fairer's 2011 essay "'Where fuming trees refresh the thirsty air': The World of Eco-Georgic" made a powerful and influential case for the ways eighteenth-century georgic poetry speaks to contemporary environmental concerns, and placed georgic at the center of literary eco-studies of the eighteenth century. The same year, John Sitter devoted a chapter of *The Cambridge Introduction to Eighteenth-Century Poetry* to "Ecological Prospects and Animal Knowledge," both reflecting a growing body of work in eighteenth-century studies and advancing the argument for the century's centrality to questions of environmental literary criticism.[25] And the year after that, Tobias Menely's award-winning *PMLA* essay "'The Present Obfuscation': Cowper's *Task* and the Time of Climate Change" brought eighteenth-century poetry and environmental history into dialogue with climate change, the Anthropocene, and the field of literary eco-studies at large. Since then, interest in exploring environmental issues in the eighteenth century has grown exponentially. In the words of the historian Alan Mikhail, the "Anthropocene has indeed become a new, eco-inflected way for us to speak about modernity and Enlightenment."[26] The work of this new scholarship has not been to scapegoat the Enlightenment in the way early ecocriticism had, however, but rather to turn to the eighteenth century with minds attuned to the ecological to better understand our current ecological moment.

Recent work by environmental historians including Paul Warde, Christophe Bonneuil, and Jean-Baptiste Fressoz has productively complicated the meaning of parallels between eighteenth- and twenty-first-century environmental concepts by reconnecting environment with politics. As Warde demonstrates in *The Invention of Sustainability*, the idea of "sustainability" qua sustainability emerges, like the usufructuary ethos, under the pressure of political and environmental changes that began to challenge older modes of engagement. Warde too locates the key driver of change at the nexus of economic and ecological transformation: "Wealth, rather than order, was becoming an overriding concern, and wealth required resources," which drove landowners to look for ways first to make land more productive, then, when signs of decline arose, for ways to make that productivity sustainable over the long term.[27] Over time, the scale of the economic and ecological problems of land management became such that they fell to regional and national governments to solve, placing sustainability and land use at the heart of the modern state. Thus it is "hard to divorce thinking about sustainability from our political organization, and it always has been," Warde writes, because the "natural" and "social" worlds have always shaped and been shaped by one another.[28]

Yet contemporary environmental discourse has tended to frame both ecological decline and human responses as modern innovations. The popular narrative of the Anthropocene depicts people in the past as stumbling into ecological disaster utterly unaware of the potential consequences of their actions.[29] Such "forgetting" is especially important to challenge, Bonneuil and Fressoz write, because it "tends to depoliticize the ecological issues of the past and thus obstructs understanding of present issues."[30] Instead, Bonneuil and Fressoz argue that "we have to think of ecology and power relations together," and that we also must historicize those relations: "If it is anachronistic to view modern societies, or certain of their actors, as 'ecological,' it is on the other hand impossible to understand their particular forms of reflexivity by envisaging them in terms of today's categories (global environment, ecosystem, biogeochemical cycles, Anthropocene, etc.), as if this offered the only valid and useful way of being 'environmentally aware.' History provides us with a space of intelligibility for grasping the localized, changing and disputed character of ways of being in the world and conceiving the place of humans within nature."[31] There has never been a time in history, Bonneuil and Fressoz argue, that human beings have not grappled with their relationship to nature; and their attempts to conceptualize their interdependence with the nonhuman world have always been embedded in and informed by their sociopolitical contexts. I pluralize "contexts" deliberately. If we are setting out to historicize the relationship of environment and politics, we must also remember that in any given moment in history we will find as great a diversity of political and environmental discourses as we do in the present.

The Usufructuary Ethos responds to Bonneuil and Fressoz's call to challenge the bifurcation of social and environmental history and to restore to the past its full, fraught, contested environmental reflexivity. By placing the usufructuary ethos within the political, social, and material contexts from which it emerges, I aim to offer not only a more comprehensive picture of an important eighteenth-century environmental ethic but also a better understanding of relationships among society, nature, and poetry. In doing so, this book also restores a thread of environmental thought missing from existing eighteenth-century environmental histories, including Warde's history of sustainability. Warde and other environmental historians focus their analyses primarily on texts like agricultural manuals and pamphlets—a logical choice for historians studying the histories of humans' material engagement with the earth, but one that fails to capture the robust discussion of human beings' relationships and moral obligations to the natural world taking place in poetry and devotional literature at the same time. Indeed, some aspects of eighteenth-century English environmental discourse that Warde takes to be missing, like the idea of "limits," or that he takes to be new, like the idea of nature being an interdependent system of circulating elements, are core aspects of the usufructuary ethos. The notion that humans' power over and right to use nature were limited, and that every creature was obligated to pass along its natural "gifts" for the support and sustenance of others, was a fundamental part of the usufructuary ethos. Those ideas were based on an understanding of the ethics of (political *and* natural) power and possession that defined their proper use by adherence to duties of support and care.

Thus *The Usufructuary Ethos* continues the scholarly project of re-evaluating the historical relationship between environment and politics, but it makes the case that without also including devotional literature and poetry, we cannot fully understand how eighteenth-century English writers understood the nature of their world or their moral obligations to nonhuman beings. As Courtney Weiss Smith has written, it takes a "willingness to take seriously the way that humans, things, God, *and* words could all contribute to the creation of meaning" to see the ways beliefs and practices were shaped by religion, politics, literature, *and* the environment.[32] The way that the usufructuary ethos merges religious, legal, political, and environmental discourses into one moral ontology that cut across social, political, and natural lines encapsulates the sort of collective creation of meaning that Smith calls for. Taking writers' claims of care for and moral obligations to nonhuman beings seriously and literally opens up a more complete picture of how seventeenth- and eighteenth-century writers understood their relationships to the environment.

That picture is not an entirely rosy one, nor is it meant to be aspirational. The usufructuary ethos situates nonhumans in moral community with human beings, but it also conceptualizes the creatures for which humans are accountable as inferior beings—beings with moral worth and significance, but beings who are nevertheless ontologically inferior to humans and thus in need of management and control for their own good. And though the usufructuary ethos was marshaled by writers who saw and deplored in the economic and colonial transformations of their time the dissolution of the moral relationship between humans and nature, it was also exploited to support and justify those very transformations. What this book offers are both the anatomy of a forgotten environmental ethic and the story of its co-option and decline.

This study traces the usufructuary ethos from the devotional and legal writings of the seventeenth century through the mercantile georgics of the mid-eighteenth century, attending to the particular political, economic, and environmental pressures that shaped, transformed, and ultimately sidelined it. The first chapter will lay out the conceptual structure of the usufructuary ethos as a way of understanding power, property, and the relationship between humans and the nonhuman world. Three main assumptions make up the structure of the usufructuary ethos: that ownership of the natural world was always displaced away from humans, who had only limited and temporary possession of it; that humans held a medial position in the hierarchy of creation; and that those foregoing facts conferred on them accountability for their uses and treatment of the nonhuman world. These three conceptual elements—displacement, mediality, and accountability—constitute the ethical assumptions that underlie the environmental thought of a diverse, numerous, and influential group of philosophical, devotional, and legal writers in the period, including Richard Allestree, author of perhaps the most popular devotional text of the long eighteenth century, *The Whole Duty of Man*; jurist and theologian Sir Matthew Hale; and John Locke. The belief that human beings are possessed of an authoritative but limited role in an inherently hierarchical world allowed writers to articulate what constituted just and unjust uses of the world and of the power gifted to humans, and to define the parameters by which "just" and "unjust" could be determined. These writers developed a deep vocabulary of terms borrowed from legal and political discourse to describe the relationships and obligations among humans, nonhumans, and God, all of which emphasize human beings' unique position of both authority and subordination in the natural order. To destroy or use up God's creation, or to flout the responsibility to preserve and care for it, violates the terms under which human beings were granted their temporary authority and their right to "enjoy and profit." Under the usufructuary ethos, human power is always limited and subordinate to a higher authority, and possession of the earth is always granted for the purposes of passing its fruits along to benefit other beings and future generations.

Yet as pervasive as the usufructuary ethos was in the late seventeenth century, it was not uncontested. The idea of humans' usufructuary ownership of the earth could, and did, support other ethical interpretations. The final part of chapter 1 turns to John Locke's *Second Treatise on Government* to uncover the tense and complex interplay between the usufructuary ethos and the economic and colonial developments of the late seventeenth and early eighteenth centuries. Locke's theory of property is premised on the usufructuary assumption that humans are God's "Workmanship" and therefore his property; and the so-called waste and spoilage provisos of the *Second Treatise* bear strong affinities to the ethical imperatives of the usufructuary ethos. Yet when Locke introduces money and a putatively empty "America" to his theory, he finds a way around the usufructuary limitations of his own theory. Locke's *Second Treatise* thus illustrates two important points. First, the usufructuary assumptions of the period did not by necessity lead to the moral conclusions I have organized under the rubric of the "usufructuary ethos." Second, the possibilities for accumulation of personal wealth and land without moral limitations or responsibilities that the interconnected phenomena of the financial revolution and colonial expansion introduced represented not only a significant challenge to the usufructuary ethos but the impetus for its prominence in late seventeenth- and early eighteenth-century discourse in the first place.

Chapter 2 begins to illustrate the way that the dominant cultural ideology of the usufructuary landlord directly shapes the forms that expressions of environmental care can take. It also shows that those expressions of environmental care were equally shaped by anxieties about the environmental failures of the landowning class, since its central texts—John Evelyn's *Silva*, Anne Finch's poetry, and John Philips's georgic poem *Cyder*—all respond in one way or another to the actual degradation of the landscapes around them, degradation they connect to changes in land use driven by desires for short-term agricultural profits. A core idea echoes through all the texts in this chapter: that nature itself embodies the intergenerational and social communities to and for which the landlord is held accountable, and that therefore the ways that the landlord uses his land carry political, moral, and environmental significance.

This is particularly true when it comes to trees. In both Evelyn's *Silva* and Finch's "Upon My Lord Winchilsea's Converting the Mount in His Garden to a Terras," trees represent organisms whose unique biological traits create both symbolic and ecologically literal connections among other beings and across generations—*if* the landlords responsible for them protect and steward them properly. My reading is a corrective to the scholarly tendency of the last several decades to read trees in early modern texts as emblems of the triumph and re-naturalizing of the Tory aristocracy and the Stuart line. That focus, I argue, has tended to reduce trees to props in the political dramas happening around them at the expense of recognizing the importance of their status as living beings. By reexamining these texts through the lens of the usufructuary ethos, this chapter also reexamines the connection between the political ideology of late seventeenth-century English writers and the nascent conservationism that has been located in Evelyn and Finch by some environmental and eighteenth-century scholars. The final section of this chapter investigates how Philips's *Cyder*—the first English adaptation (as opposed to translation) of the georgic—merges the genre's classical authority and ethos of forward-looking care with the usufructuary ethos's concerns with environmental duties to public and posterity and the ethical questions of the use of riches. By reading *Cyder* through the lens of the usufructuary landlord, the final section of this chapter situates the georgic within the usufructuary ethos and shows that the poem's political and environmental points of view are inextricable from one another.

Chapter 3 charts new tensions as the inherent weaknesses and limitations of the usufructuary ethos emerge under the pressures of the political, economic, and environmental changes of Walpole's England. Pope's satires of the 1730s betray his anxiety—and nascent acknowledgment—that consolidating property and power in the hands of one group leaves public and posterity vulnerable to their whims. Pope's *Horatian Imitations* and *Epistles* to *Burlington* and *Bathurst* attest to both the continued influence of the usufructuary ethos and the growing awareness of its failures. Pope's critiques of gentlemanly misuse and abuse of property and power participate in the longstanding series of similar critiques discussed in previous chapters, which together reflect an anxiety about the difficulty, if not impossibility, of controlling the choices made by the landlord class, particularly in the face of economic, social, and environmental developments that re-created the worlds the English inhabited. Given the rapidly expanding development of paper credit, which many perceived to be fundamentally altering the relationship between human beings and their world, and the ways such forces contributed to the transformation of the English landscape, the centrality of individual landlords' choices about what they did with their possessions took on greater social and moral weight. Pope sees and draws attention to this, but also recognizes the problem it poses for the effectiveness of the established usufructuary ethic of use. Moral admonishment is a weak bulwark against misuse. The consistent skepticism Pope evinces about the choices of those in power in England, his insistence upon individual choice and retreat as the best feasible solution to the problems he sees, and his consistent and vivid portrayals of the social and environmental decay brought on by individual misuse all indicate the depth and sincerity of his awareness of and anxiety about this widening crack in the moral and practical power of the usufructuary ethos.

Chapter 4 highlights a fundamental shift in the English georgic that took place between the 1690s and the mid-eighteenth century, one that both reflected and was driven by shifts in the material and conceptual relationships among humans and the environment as a result of the expansion of mercantile colonialism. This chapter maps the confluences of the georgic and the usufructuary ethos through two key eighteenth-century English georgic poems—John Dyer's 1757 *The Fleece* and James Grainger's 1764 *The Sugar-Cane*—to demonstrate that the economic and material changes English agriculture was undergoing during the eighteenth century, which georgics so famously dug into, reshaped the usufructuary ethos. The midcentury mercantile monocultural georgics studied in this chapter reflect the increasing tendency, foreshadowed in previous chapters in the anxieties of writers like Pope, to understand the environment as wealth-generating, rather than sustaining; as a site for transformation in the service of a key product, rather than of sustenance for humans and nonhumans. Yet that poets continued to turn to the georgic mode suggests that they desired to see themselves and the transformations they depicted as in continuity with the past, as perpetuating the usufructuary ethics of use that still carried deep cultural authority and informed the understanding of human-environment relationships.

That "dialectic of conservation and expansion" is typical of the deep ambivalence Suvir Kaul uncovers at the heart of eighteenth-century poetic defenses of commercial expansion, which suggest "continuity as much as contrast—the link between the past and the future, the complication of forward-looking poetic desire by forms of historical nostalgia," not least of which is nostalgia for traditional, usufructuary bonds of landlordship and obligation that the classical authority of the Virgilian georgic had helped bolster.[33] Specifically, *Fleece* and *Sugar-Cane* attempt to preserve continuity with older values by transferring the locus of socio-environmental connection and value from land and trees to trade and commerce. I argue that shifting the work of social cohesion from the environment to the economy represents a fundamental change both from earlier instantiations of the georgic and from the usufructuary ethos. Dyer and Grainger apply the language and topoi of the usufructuary ethos and the georgic to enclosure, mercantile monoculture, and colonialism in order to create the sense that nothing fundamental has changed in English values or methods—that the pursuit of wealth, which both poems, but particularly Grainger's, acknowledge is the goal of such work, is not only permissible but positive. While neither breaks entirely with the caution toward use of riches seen in chapter 3, both work to square the pursuit of wealth qua wealth with usufructuary values. *The Fleece* and *The Sugar-Cane* offer especially clear examples of how the material and economic changes of the latter half of the eighteenth century eroded the usufructuary ethos: not by attacking or discrediting it directly, but by incrementally, even unintentionally, transforming it by using its forms and rhetoric to soften and elide the profundity of the changes taking place.

1

The Usufructuary Ethos

Roots and Branches

IN 1682 the popular Dissenting preacher John Howe published a sermon in which he described the relationships among God, humans, and nonhuman beings: "God indeed is the only *Proprietor*, Men are but *usufructuaries*. They have the use of what his providence allots them; He reserves to himself the *property*; and limits the *use so far*, as that all are to be accountable to him for all they possess. And are to use nothing they have, but as under him, and for him, as also they are to do themselves."[1] Howe's definition of God as the "only *Proprietor*" and humans as "but *usufructuaries*" captures the fundamental premise of the *usufructuary ethos*: that human beings are not the true owners of the nonhuman world, in either a spiritual or a literal sense, and that their status as subordinate possessors of the natural world imposes ethical responsibilities on their use of it. To name humans the "usufructuaries" of God's natural world is to place them implicitly under his absolute authority for ensuring the preservation of the land for the benefit of other beings and future generations.

Howe's sermon provides a helpful breakdown of the structure and implications of the usufructuary ethos. The way that usufruct displaces proprietorship (God's) from the right of use (humans') enables Howe to reinforce to his readers that their existence and subsistence come with moral obligations. Human beings are positioned in between God and creation; they use what God has temporarily granted them "under him, and for him," a qualification which emphasizes the fact that human authority is predicated on the persistence of God's authority over his creation. Neither human possession nor human power is absolute. In order to enforce compliance with those limitations, Howe says that humans are "accountable" to God for the property they possess, as well as for ensuring that the uses to which they put it conform to God's guidelines. Deploying the legal term "usufruct" allows Howe to articulate the boundaries of just and legitimate use of human authority and power over the nonhuman world, and to remind his audience that what humans *can* do with God's gift of creation and what they *ought* to do are two different things. Howe's description of the usufructuary relationship among God, humans, and nature in this sermon echoes numerous late seventeenth- and early eighteenth-century writers from diverse political and theological backgrounds, all of whom share with him both the vocabulary of usufruct and the belief in the usufructuary nature of human possession and its ethical ramifications.

This chapter will lay out the three core principles of the usufructuary ethos as they are articulated in the philosophical, devotional, and legal writing of the late seventeenth and early eighteenth centuries: (1) That the human right to use nonhuman creation is *displaced* from absolute ownership of it, which rests permanently with God; (2) that human beings have a *medial* relationship to power and to the nonhuman creation, in that they are both under and in authority, and are the media through which the means of subsistence pass from organism to organism and from one generation to the next; and (3) that human beings are *accountable* for the care of the earth and its creatures both to the absolute proprietor and to future generations of users. By tracing these principles through the works of influential philosophical, theological, and legal writers, I will demonstrate that the usufructuary assumptions of displacement, mediality, and accountability of human property and power were a fundamental part of ethical thought during the period, and that they are key to understanding the ways that English writers of the late seventeenth and early eighteenth centuries related to the nonhuman world. Borrowing language from the legal concept of usufruct to distinguish God's absolute dominion over creation from mankind's usufructuary dominion offered a way for late seventeenth- and early eighteenth-century writers to persuade readers that humans were morally accountable for the care and preservation of the earth and its creatures.

Furthermore, "usufruct" appears in seventeenth- and eighteenth-century writing with clusters of other terms: *steward, tenant, landlord, viceroy*. Like *usufructuary*, each of these terms refers to a person on the middle rungs of a hierarchy, imbued with authority but also subject to it. Likewise, they are terms of legal, social, and political power that the texts examined in this chapter apply to relationships among nonhuman beings and God as well as humans. Combined with the usufructuary ethos's core principles, these linguistic patterns reveal that the usufructuary ethos subtended understandings of power and possession across social and environmental lines. It provided an ethical foundation for the hierarchical relationships most early modern English writers believed existed among living beings, human and nonhuman alike.

Yet the usufructuary ethos was not uncontested. The final part of this chapter turns to Locke's *Second Treatise on Government* to uncover the tense and complex interplay between the usufructuary ethos and the economic and colonial developments of the late seventeenth and early eighteenth centuries. Locke's theory of property was grounded in the assumption of usufructuary possession, and the so-called waste and spoilage provisos bear strong affinities to the ethical imperatives of the usufructuary ethos. But by introducing money and a putatively empty "America" to his theory, Locke finds a way around the usufructuary limitations of his own theory. Locke's *Second Treatise* thus illustrates two important points. First, the usufructuary assumptions of the period did not by necessity lead to the moral conclusions I have organized under the rubric of the "usufructuary ethos." Second, the possibilities for accumulation of personal wealth and land without moral limitations or responsibilities that the interconnected phenomena of the financial revolution and colonial expansion introduced represented not only a significant challenge to the usufructuary ethos but the impetus for its prominence in late seventeenth- and early eighteenth-century discourse in the first place.

But first, we turn to the history of the core concept of the usufructuary ethos: displacement.

Usufructuary Displacement and Natural Law

The core premise of the usufructuary ethos, "displacement," can be traced back to the thirteenth-century philosopher and theologian Thomas Aquinas. The foundational text of natural law theory, Aquinas's *Summa Theologica* argues for the existence of moral rules that "rational" beings can discern from nature without the necessity of divine revelation or positive law.[2] Those "natural" laws would represent the manifestation of God's moral will in the universe he created. Aquinas argued that because natural law emanates from and reflects the divine, God could not break it. But in order to make his case that moral laws against acts like murder and theft were universally, eternally binding, Aquinas had to deal with the fact that the Bible depicts God himself killing people and taking their belongings. If, as Aquinas claims, God "can do nothing outside the natural order," how can laws he breaks be said to be part of the natural order?[3] Usufructuary displacement solves that problem: "whatever is taken by the command of God, *to Whom all things belong,* is not taken against the will of its owner, whereas it is in this that theft consists. Nor is it only in human things, that whatever is commanded by God is right; but also in natural things, whatever is done by God, is, in some way, natural."[4] Because everything belongs permanently and absolutely to God—including the lives of the beings he creates—God can never be said to be taking or destroying something that he did not have the right to take. Aquinas had ample biblical support for the view that all human possession—even of life itself—is fundamentally usufructuary; see, for example, Leviticus 25:23: "The land shall not be sold forever: for the land is mine; for ye are strangers and sojourners with me."[5] God the proprietor's rights always supersede those of the creatures to whom he grants the usufruct of their lives and goods.

By the seventeenth century, the notion of God's permanent ownership of his creations had become the philosophical basis for the bindingness of any law, whether natural, ecclesiastical, or common. That migration is unsurprising, given that natural law dwells in the overlap between ethics, law, and religion, often combining all three. This was particularly true in the early modern period, when the lines dividing moral philosophy from theology from law were less deeply drawn than they are today. Displacement served as one of the cornerstones of the Dutch natural lawyer Hugo Grotius's *De Jure Belli ac Pacis* (*The Rights of War and Peace*) (1625), arguably the first European treatise of modern international law. Even in cases where the law of nature seems to be have been breached by God's own commands to kill someone or to take something, Grotius writes, it "always remains the same, is not changed; . . . that cannot be Murder or Theft, which is done by the express Command of him who is the Sovereign Lord of our Lives and Estates."[6] Natural law's universality and permanence, Grotius reasoned, provided a set of laws that bound all people, regardless of their local laws or customs, making it possible to conceive of laws that could bind not only subjects of a particular sovereign but also the relations between sovereigns and their nations.

In England, the philosopher and theologian Edward Stillingfleet used God's "absolute dominion" to argue in his 1659 *Irenicum* that the rules governing different churches could be both arbitrary (that is, chosen by individual churches rather than prescribed by the Bible) *and* binding, providing grounds for Latitudinarian toleration of Presbyterianism by Anglicans.[7] For Stillingfleet, the fact that the "Law of Nature immediately binds the soul and conscience of man" provides the basis for the bindingness of positive covenants made between humans and God and between humans and other humans, and that is what allows congregations to covenant together as communities of faith.[8] God's superior dominion likewise provided the basis for the validity of human laws in analyses of English common law, most famously in William Blackstone's 1765–66 *Commentaries on the Laws of England*.[9] Blackstone's use of displacement in "Of the Nature of Laws in General" reflects a long tradition of invocations of the concept in English discussions of both civil and common law dating back to at least the sixteenth century.

Displacement became the cornerstone of another set of moral arguments in the seventeenth century as well: those concerning the possession and use of the nonhuman world by human beings. It was the common starting point of arguments about the ethical limitations of human rights of use among both philosophers and theologians. For Samuel Pufendorf, the German jurist and philosopher who was one of Locke's chief influences, displacement provided the foundation for the ethics of the "power of mankind over the things of the world": it is "beyond Dispute, that Almighty God" has an "originary and super-eminent Property over all," while the "Right which Men hold over Things [is] *Usufructuary* only," Pufendorf wrote in *De Jure Naturae et Gentium* (*The Law of Nature and Nations*) (1672).[10] Though Pufendorf denies that there is a "moral community" among humans and nonhuman beings, he argues that to consume nonhuman beings "idly or wantonly" would be to "Dishonour" God, the "Giver of so great Gifts."[11] Locke uses similar logic to distinguish a "*State of Liberty*" from a "*State of License*" in the *Second Treatise on Government*, arguing that "no one ought to harm another in his Life, Health, Liberty, or Possessions" on the basis that all beings are God's property, not their own: "For Men being all the Workmanship of one Omnipotent, and infinitely wise Maker; . . . they are his Property, whose Workmanship they are, made to last during his, not one anothers Pleasure."[12] Locke's invocation of displacement in the *Second Treatise* reflects his recapitulation in his unpublished *Essays on the Law of Nature* of the same usufructuary argument for the bindingness of natural law discussed by his forebears Aquinas, Grotius, and Pufendorf.[13]

For all those similarities, major differences emerge when we examine in more detail the ways different writers interpreted the moral ramifications of displacement for the relationships between humans and the nonhuman world. Stillingfleet, focused on the issue of church government, does not mention nonhuman beings at all. Both Pufendorf and Locke denied that human beings were in a "moral community" with nonhuman beings, but they diverged on the subject of whether humans had any moral obligations governing their treatment of nonhumans. While Pufendorf still insists on an obligation to God not to use his creatures "idly or wantonly," Locke abandons even that limitation. For him, the obligation is entirely to other humans: a person is bound to use resources in a way that allows him "as much as he can, *to preserve the rest of Mankind*," for humans are not "made for one anothers uses, as the inferior ranks of Creatures are for ours."[14] For Locke, displacement provides the basis to argue that it violates other *humans'* rights to hoard or destroy resources—hence the *Second Treatise's* two "provisos" against spoilage and waste—but the nonhuman resources themselves, be they animal, vegetable, or mineral, are not a part of the moral equation.

In stark contrast to Pufendorf's and Locke's denial of "moral community" with nonhuman beings, devotional writers of the seventeenth century frequently used human beings' usufructuary possession of the earth to explain their moral obligations not only to God the possessor but to the nonhuman beings he created as well. These authors insisted upon the distinction between absolute and usufructuary possession in order to articulate a distinct environmental ethic of mediality and accountability, built upon the complex dynamics of the hierarchical power disparity they believed existed among God, humans, and nonhumans. These ideas form the core of the usufructuary ethos, and to these I will now turn.

MIDDLE MAN: USUFRUCT AND HIERARCHY IN SEVENTEENTH-CENTURY DEVOTIONAL WRITING

In the introduction to this chapter, we have already seen one example of the usufructuary ethos laid out by leading Nonconformist minister John Howe, who stated that "God indeed is the only *Proprietor,* Men are but *usufructuaries* . . . [and] are to use nothing they have, but as under him, and for him, as also they are to do themselves." Implicit in the second half of Howe's statement is the principle that the rights of human use and possession—of their own bodies as well as of the earth—were bound by the will and rules laid down by the absolute proprietor, God. Whereas Pufendorf claimed that to use God's creatures "idly or wantonly" would "dishonor" him, and Locke dismissed nonhumans altogether, Howe interprets humans' usufructuary status to mean that their use of the nonhuman world comes with explicit limits: "It is indeed the greatest absurdity imaginable, that they who are not Masters of themselves should think it permitted them, to use what comes into their hands, as they list; for the service of their own lusts, and the gratifying of a rebel flesh. . . . Or that he who hath so absolute a right in them, should not have that right in what he hath committed to them, as to prescribe rules to them, by which to use and imploy it."[15] Howe is reminding his readers of their medial position between God and creation, under the authority of the former, who dictates the terms of their authority over the latter. By their very nature, Howe argues, human beings are in the middle of the hierarchy of creation. With that unique position comes a number of moral obligations, the most important of which is to obey the rules God set down for the use of his gifts, which explicitly forbid uses that serve human desires and appetites without reference to God's will and purpose in creating the world. The importance of Howe's attribution of moral significance to the things humans do to and with the nonhuman world should not be understated. The medial position he describes for humans puts them in a moral relationship with nonhuman creatures, extending to nonhumans a place in the moral universe denied to them by natural law philosophers like Pufendorf and Locke. For Howe, the "moral community" does indeed include nonhuman creatures, and usufructuary man's moral obligations extend to them.

The fact that it was Howe who made that argument is also important. Howe was one of the most highly respected Nonconformists in England in the second half of the seventeenth century. He was consulted by both James II and William of Orange on matters of religion and toleration.[16] When *Self-Dedication* was published in 1692, Howe was enjoying the height of his London ministry's popularity as well as his most prolific period of publication. That alone suggests a large and powerful audience for Howe's declaration of humankind's usufructuary mediality. Nor was Howe alone in making such a declaration. Ministers and theological writers of all stripes and statuses, Anglican and Dissenting, popular and little known, invoked the same idea in similar language, demonstrating how commonplace an assumption it was that human beings held a medial position as God's usufructuaries. The most influential Anglican among these writers was undoubtedly Richard Allestree, who elaborated on man's usufructuary mediality and its moral consequences in almost all his written works. Allestree (1621/2–81) was a royalist whose most famous work, *The Whole Duty of Man*, "took its place with the Bible and the Book of Common Prayer as an indispensable possession in every respectable household" from its publication in 1657 through the end of the eighteenth century.[17] References to *Whole Duty* appear in eighteenth-century texts ranging from *Shamela* to *The Rivals*, and in the correspondence of figures including Jonathan Swift and Benjamin Franklin. The unprecedented popularity of *Whole Duty*[18] extended to Allestree's other devotional writings in the eighteenth century as well; additional texts attributed to Allestree, including *The Art of Contentment* and *The Gentleman's Calling*, were published with *Whole Duty* under Allestree's name starting in 1684, and were near constantly reprinted throughout the eighteenth century.

Allestree's most straightforward articulation of the usufructuary principle appears in 1675's *The Art of Contentment*, in which he states that the ancient Israelites "were but [God's] *usufructuaries*: and tis as evident that our *tenure* is but the same."[19] But in his 1660 guide to the moral obligations of the English landlord, *The Gentleman's Calling*, Allestree had a more complex point to make about usufructuary mediality: that every single part of creation, animate and inanimate, human and nonhuman, acts as a medium for God's "gifts," passing them from one being to the next. If a person "direct his eyes upward on ... the glorious Spirits that encompass God's Throne," Allestree wrote, "he will not in all the Records of Earth or Heaven find ... any Clause of Exception in this universal Law" requiring creatures to perform their duties to other creatures as prescribed by God.[20] Angels and animals serve equally as exemplars for human beings, and are as like each other, in this sense, as human beings are to either of them. The "common obligation" to God across every scale of being makes nonhuman beings not merely objects of human stewardship but living exemplars of the universal moral duty to uphold God's order by fulfilling one's place in it. Like "the Earth which though she conveys the springs through her veins, yet is allowed to suck in so much, as may give her competent refreshment," so the gentleman may enjoy the fruits of the property God grants him, so long as he fulfills the broader ends for which God granted him his power and riches.[21] Allestree drives home the seriousness of the gentleman's duty by showing that it is of a piece with the very moral fabric of the created world. While the role of the usufructuary of the earth was specifically humankind's, *having* a position in the order of creation that comes with rights and duties was universal, a necessary piece of being part of the "divine Hierarchy" that makes up the cosmos.[22]

For seventeenth-century theological writers, then, usufructuary mediality entailed a couple of important moral conclusions. One was the necessity of human humility: the fact that God had given all of his creatures a role to play, different from humans' but equally important, undermines any notion that human beings are categorically different from nonhuman beings. The other was human beings' obligation faithfully to perform the duties requisite to their role for the well-being of all. Thomas Adams, another popular Anglican preacher active in London earlier in the seventeenth century,[23] illustrates both these points in his description of the roles of the earth, the creatures, and "usufructuary" man in a 1623 sermon: "The earth, in a thankefull imitation of the Heauens, lockes not vp her treasures within her owne Coffers; but without respect of her priuate benefit, is liberall of her allowance, yeelding her fatnesse and riches to innumerable creatures, that hang on her breasts, and depend vpon her as their common mother for maintenance: Of the beasts that feed vpon her, Kine giue vs their milke, Sheepe their wooll: euery one payes a tribute to man, their vsufructuary Lord: this is their *fruites*."[24] Humans are designated here as God's usufructuaries, the recipients of the "fruites" of the rest of creation, but not the outright owners of the earth or its fruits. The "barren tree" of Adams's title refers to human beings as God's usufructuary creations: "God is our Lord and Proprietary, England is his Vineyard, every one of us his Fig-tree, thus planted, watered, blessed by his gracious mercy."[25] Adams's purpose in invoking the role of the earth in giving succor to sheep and cattle, and sheep and cattle their "tributes" to humans, is to reinforce humans' responsibility to fulfill their duty to pass their "tribute" along as other creatures do. The rhetoric of "tributes" Adams uses seems at first blush to emphasize human beings' superiority to the rest of creation, but his main point is actually that human beings can glean the existence of their duties as medial beings from observation of nature.

Adams's sermon offers one final hint that will remain crucial in all future discussions of the usufructuary lord. The fact that the earth "is liberall of her allowance," rather than "lock[ing] vp her treasures within her owne Coffers" emphasizes the point that God's gift of "riches" was meant to benefit more than merely the recipient of them. Human beings—particularly the wealthy and landed—were granted possession of these gifts in order to pass them along to others, in one form or another, just as every other being God created has the duty to pass the benefits of their gifts to other living beings. Allestree repeated this lesson throughout his writings, as well. In *The Gentleman's Calling*, Allestree exhorts the landlord not to "give up his ear to be bored by Mammon, when God proclaims a Jubilee," declaring that "if he leave" the "indigent parts of [God's] Family . . . destitute, and suffer either his riot or his covetousness to feed upon their Portions, what more detestable falseness can be committed, not only in respect of them, whose right he thus invades, but of God also, whose trust he abuses?"[26] In *The Whole Duty of Man*, he further insisted that it is man's duty "to distribute" his gifts "to them that want; and therefore not to do it, is the same injustice and fraud, that it would be in any Steward to purse up that Money for his private benefit, which was intrusted to him, for the maintenance of the Family."[27]

Writers like Adams and Allestree perform a delicate balancing act, weighing human beings' uniquely extensive right to use and benefit from the "fruites" of earthly creation against their obligation to understand their possessions as theirs only temporarily and partially, given only in order to be distributed. Usufructuary mediality thus places human beings in an interdependent, yet still hierarchical, interspecies community with all other created beings in which the material goods of creation and the basis of subsistence flow up and down the chain from being to being. Human beings occupy a place "above" many others in the chain of being, but this thread of usufructuary mediality emphasizes the obligations human beings have in common with the rest of creation, rather than their exceptionalism.

Power, Mediality, and Legal Metaphors: Sir Matthew Hale and the "Vice-Roy of the Great God"

In the final example from *Whole Duty* in the last section, Allestree invoked a series of legal metaphors to explain human beings' obligation to distribute their gifts. The gentleman is a "Steward," who "invades the rights" of those looking to him for support, "abuses God's trust," and commits "injustice and fraud" if he keeps his goods to himself. The term "tenure" he used in *Art of Contentment*—the biblical Israelites "were but [God's] usufructuaries: and tis as evident that our *tenure* is but the same"—is an English term for the "title by which the property is held" and the "relations, rights, and duties of the tenant to the landlord."[28] John Hales, a former prebendary of Windsor and professor of Greek at Oxford in the mid-seventeenth century, piled on even more legal metaphors in a sermon first published in 1660: "When we call the things of the world *ours*, ours is but a word of usurpation; we peradventure may be some *emphyteuticaries*, or *farmers*, or *usufructuaries*; but the propriety is in another person," that other person being God.[29] In this case, Hales links usufruct to another Roman legal concept, *emphyteusis*, which refers to the "right to property held under a long-term or perpetual lease."[30] Given that both Roman terms emphasize humans' limited tenure, Hales most likely invokes the English term "farmer" not in the modern sense of one who grows food, but in the older sense of one who rents land to cultivate either for himself or for an owner.[31] Like usufructuary and emphyteuticary, then, "farmer" in this context connotes displaced ownership and limited rights of possession and use.

All of these legal terms refer to relationships defined by property. More specifically, they are legal terms that define relationships between parties according to their relative degrees of *power* over the use of the earth. The examples in the previous section emphasized the relative lack of power the usufructuary has compared to the proprietor, but in other places writers used these legal metaphors to contemplate the moral significance of humans' position of greater power over the rest of creation. This perspective is more focused on humans' unique powers and rights than their subjection to a higher authority, but for the explicit purpose of delineating the boundaries that separate just from unjust uses of that power. The ethical boundaries of power were predicated on the same sort of usufructuary medial logic as the ethical limits of possession and use adumbrated in the last section. The question, as ever, was what humans owed not only God but also the creatures "below" them in the natural hierarchy.

Probably the most important late seventeenth-century analysis of human usufructuary power comes from Sir Matthew Hale. Hale is best known as the author of *Historia Placitorum Coronae* and *History and Analysis of the Common Laws of England*, the former of which served as the main book on English criminal law in the eighteenth century, and the latter of which provided Blackstone with the structure upon which he based his own *Analyses*. Almost all of Hale's writings were published after his death in 1676, including his theological magnum opus, *The Primitive*

Origination of Mankind, which considered at length human beings' moral obligations to each other and to the rest of creation. Owing partly to his posthumous publications and partly to his reputation for personal integrity and piety, in the eighteenth century Hale came to be considered a moral exemplar by people of all religious and political allegiances. His toleration and the development of his theology in *Primitive Origination* were major factors in that reputation. Hale was sympathetic to Noncon-formists before and after the Restoration; he famously refused to pros-ecute a Quaker meeting for violating the statute against conventicles in the early 1660s. Though he remained a practicing Anglican throughout his life, his religious views were not dictated by any single denomination. In the later 1660s, his biographer Alan Cromartie explains, his "broadly ethical religion" had broadened enough that he even "expected virtuous pagans to be saved," a position, among others, that Hale developed as a part of his personal quest for theological and moral truth.[32] His posthu-mous reputation helped drive continued interest in his religious writings, which were reprinted throughout the eighteenth century.

Hale's devotional writing epitomizes the ways that legal, theological, and moral discourses intersected and reinforced each other in the usu-fructuary ethos. Attending closely to a key example from *Primitive Origi-nation* in which Hale juxtaposes language of human quasi sovereignty over creation with language of obligation will clarify the nature of human beings' usufructuary medial position of power, and the way that position shaped human obligations not only to God but to the rest of creation. Hale defines a broader—though still fundamentally limited—role for human beings than Howe had in *Self-Dedication*, emphasizing their role as God's agents on earth rather than their subjection to God's will and power. Hale chooses metaphors to describe humans that emphasize their relative authority, stating that God made humankind the "Vice-Roy of the great God of Heaven and Earth in this inferior World; his Steward, *Villicus*, Bayliff or Farmer of this goodly Farm of the lower World, and reserved to himself the supreme Dominion, and the Tribute of Fidel-ity, Obedience, and Gratitude, as the greatest Recognition or Rent for the same, making his Usufructuary of this inferior World to husband and order it, and enjoy the Fruits thereof with sobriety, moderation, and thankfulness."[33] Defining humankind's role first as "Vice-Roy" implies sovereignty, and with it comes a degree of power and autonomy also ab-sent from previous examples. Hale goes on to call humankind the "Vice-regent of Almighty God, in the subordinate Regiment especially of the

Animal and Vegetable Provinces," and he claims that God made humans his vice-regents because the "Earth, and Vegetables, and Animals stand in need of such a Superior Nature to keep them in a competent order." Without the careful protection of man, "mansuete Animals" and plants would fall "prey of savage Beasts," destroying the delicate order God intended for his creation. These are all familiar early modern arguments for human control over nature.

Yet the terms with which Hale parallels "vice-roy" and "viceregent" all refer to people whose duty it is to care for that which belongs to another, and to carry out orders and justice dictated by a higher authority. A "*villicus*" was a Roman steward, charged with carrying out his overlord's orders and managing the lord's farms and estates.[34] A bailiff executes decisions made by a higher court, and a steward tends to the well-being of a lord's estate and bears a fiduciary duty to protect its best interests. Each of these metaphorical titles illustrates the nature of the power granted to human beings in the usufructuary ethos. They are the medium through which God's will is enacted.

Furthermore, Hale is careful to couch his description of human power in terms of the divine purpose for which that power was conferred. Despite granting humans extensive control over nonhuman creatures, Hale insists that God gave humans only a "*subordinate* dominion." The power Hale's "viceregents" have been granted is fundamentally limited by the requirement to "husband" as well as "order," and to do so with "sobriety, moderation, and thankfulness." Human authority, therefore, is dependent for its legitimacy on the fulfillment of particular terms. On the one hand, humans have a kind of sovereignty over creation, but on the other, that sovereignty is explicitly limited and revocable. It is subject to the higher sovereignty of God, as well as to human beings' performance of their obligations to both God and nature. Hale's perspective looks down the chain of being at the rest of creation, but he defines human beings as creatures responsible to God for other beings. Though anthropocentric, the belief that humans are God's agents on earth did *not*, in the usufructuary ethos, imply an untrammeled right to exploit or plunder the earth. Rather, as a creature that exists in between God and the rest of creation, and one whose rights and powers are mediated by those moral relationships, humans have an obligation to carry out the duties set forth for them by their creator for the good of the rest of creation.

In another of his devotional writings, first published in 1676 as *Contemplations Moral and Divine*, Hale further develops a vocabulary of

terms that carried simultaneously legal, political, and religious connotations in order to refute the idea that "in lending" human beings his
"own Creatures for [their] Use,"[35] God was giving them free reign over
his creatures. Hale's discussion of human beings' relationship to nonhumans appears in a section of *Contemplations* entitled "The Great
Audit, with the Account of the Good Steward," which was excerpted,
condensed, and reprinted regularly as a stand-alone pamphlet from the
1690s through the 1790s.[36] Seven separate editions of *The Great Audit*
appear in *Eighteenth Century Collections Online (ECCO)* dated between
1766 and 1800 under the titles *An Epitome of Judge Hale's Contemplations,
in his Account of the Good Steward* and *The Great Audit, or Good Steward,
Being some Necessary and Important Considerations to be considered by all
Sorts of People, taken out of the writings of the late Worthy and Renowned
Sir Matthew Hale*. A 1788 edition of the latter title is marked as "the thir-
teenth edition," evidence of even greater popularity and circulation of
the pamphlet than the limited *ECCO* sample evinces. The emphasis on
Hale's identity as the author in both titles indicates his status as a moral
authority in the minds of the eighteenth-century English reading public.
That *The Great Audit* was labeled an "epitome" of Hale's moral writing
further establishes that this remained the most relevant aspect of Hale's
moral thought to readers in the eighteenth century.

    The Great Audit details the responsibilities that must be fulfilled in
order to make a person a "good steward" of God's gifts. In the subsection
titled "Touching Thy Creatures"—the longest of the eight subsections
included in the condensed eighteenth-century editions—Hale emphasizes humankind's medial position between God and nonhumans and
their role as "dispensers" of God's gifts: "I rejoiced in this, that my Master
esteemed me faithful, committing the Dispensation [of the earth] to my
Trust; but I thought it no more mine, than the Lord's Bailiff, or the merchant's cash-keeper, thinks his Master's rents or money his: And therefore thought it would be a Breach of my Trust to consume or embezzle
that Wealth in excessive Superfluities of Meat, Drink, or Apparel, or in
advancing to myself or my Posterity to a massy or huge Acquest."[37] The
fiduciary language of trust, debt, abuse, and even embezzlement Hale
uses to describe either greedy or avaricious abuse of the nonhuman
world echoes that of Allestree. Both writers make the same point: to be
in possession of power and riches is to be entrusted with the proper *distribution* of those riches; keeping them for one's self or one's family violates the terms of the gift. Most importantly, fiduciary metaphors such

as these made it possible to articulate the stakes of flouting one's duty to God and other beings. It is a violation, a "breach." It is *unjust*. It repositions the reader in relation to his power: imagine your bailiff or your cash-keeper stole from you, Hale suggests; that is what you are doing when you keep your riches to yourself, rather than using them for the good of all. Just as the bailiff's breach sows social discord and disorder on the estate, so the landlord's breach sows ontological discord and disorder in God's creation.

Thus, though Howe, Adams, Allestree, and Hale invoke human beings' usufructuary mediality in different contexts and to different ends, together their texts point to an underlying assumption that the usufructuary structure of limited, medial possession and power is the building block of human relationships to the divine and to creation. Usufructuary mediality makes it possible to emphasize human beings' subjection to a higher authority without ignoring or erasing their power, and conversely, to emphasize the fact that humans possess power and authority without erasing the existence of limitations to the exercise of that power. Metaphors of political sovereignty (viceroy, vice-regent) and legal power (usufructuary, villicus, bailiff, steward) made it possible to articulate a notion of limited, conditional sovereignty that could be either justly or unjustly exercised, and to argue that those distinctions depend upon how the use of that power—and the use of the nonhuman beings over which that power extends—affects others, both in the present and in the future. The deep contemplation of usufructuary mediality and its moral importance to human-nonhuman relationships demonstrated by such a prominent and influential lawyer and judge as Hale points to how deeply intertwined the usufructuary ethos was in theological, moral, political, legal, and environmental thought.

Usufructuary Mediality and the English Landlord

One final set of legal metaphors, derived from English landlord-tenant relationships, elucidates the ways that the usufructuary ethos intersected with the social hierarchies of early modern England. One good example of these ideas coming together in a popular religious text comes from Anthony Horneck, a prolific Anglican clergyman in the late seventeenth century, whose crowds of London parishioners Richard Kidder once asserted stretched beyond the capacity of his church to contain.[38] A close friend of Gilbert Burnet's and a member of the Royal Society, Horneck's devotional writings were reprinted consistently through the 1730s. In his 1686 *The First Fruits of Reason*, Horneck described the relationships among God, humans, and nonhumans in terms of both the proprietor-usufructuary relationship and the English landlord-tenant relationship: "He that creates gives all that the Creature hath; and it's hard, if he that makes the Tenant and gives him Lands and Houses, may not reserve to himself a quit-rent, or a Pepper-corn rather, as an acknowledgement, that the Creature is the Usufructuary of his possessions. All the service man can do, or that God requires of us, is nothing but a small and inconsiderable Rent, our great Landlord reserves, whereby we may own him the Maker and Author of our welfare."[39] What the usufruct and tenant metaphors Horneck juxtaposes have in common is that they both place the human in a medial position between creator and creation, subject of and accountable to the lord for the use of the gift of creation he has placed under human power. Like Howe, Horneck invokes the legal metaphor of usufruct for humankind in order to drive home a moral lesson about the necessity of humility and thankfulness to God, exhorting his readers to think about and give thanks to the God who created everything they have, rather than simply taking it as their due. Yet the addition of the landlord-tenant metaphor invokes additional layers in the relationships and obligations that exist between God, humans, and creation. Without overtly stating it, Horneck's addition of a landlord and tenant metaphor suggests that the same ethical structures subtend *all* relationships between humans and nonhuman creation, whether religious, social, or political in nature.

Specifically, the relationship of medial power between God the proprietor and humans the usufructuaries is mirrored in the social realm in the form of the landlord and tenant. The God/landlord parallel enables Horneck to make the further claim that the religious duty of thankfulness to God is an issue of *fairness* to him as the ultimate proprietor/creator—what humans owe him as "rent" in exchange for what God has granted his "Creatures"—and implies that humans are bound in obligation to God by the same principles of equity that bind them to their legal obligations in the political realm. This has the dual effect of giving divine justification to the political relations of property ownership in England, while also linking the justification of those relations of property ownership to specific moral obligations and limitations: God is the divine landlord and all of creation are his tenants, who rely upon God's will for the means of their subsistence. Furthermore, not only does the metaphor of landlord and tenant indicate that humans' moral obligations extend to their treatment of the nonhuman world (the "Lands" God grants them for their "welfare"), but Horneck's use of the term "Creature" rather than "man" or "person" to describe the receivers of God's gifts establishes that these are universal laws of obligation, applicable to humans and nonhumans alike. It is human beings' *createdness* that places them in the relation of a tenant to their landlord God; they share that createdness, and therefore that basic relationship, with the rest of God's creations. In this sense, while humans are certainly special in Horneck's view, and are given unique rights and roles by God, they are, as we saw earlier in Adams and Allestree, different from other creatures only in degree, not in kind. All creatures are governed by the same originary (we might say natural) laws, and accountable to the same creator. Even in texts like Horneck's, where the moral focus is not explicitly environmental, the environmental relevance of the usufructuary relationship he lays out is strongly implied in the language he chooses. The basis for an environmental ethos was implicit in the beliefs and assumptions that structured the usufructuary ethos, therefore, even in writings where that context was not the focus.

The deep connections between the figure of the English landlord and the usufructuary ethos will be discussed in greater detail in the following chapter. For now, what's most important about Horneck's example is the fact that he focuses on the way the usufructuary ethos places humankind under the authority of God, emphasizing their lack of power and authority relative to their lord, and the fact that Horneck links the moral and ethical duties incumbent upon human beings as human beings to those incumbent upon them as members of a particular society. For Horneck's purposes—persuading his reader to feel and act on the proper sense of gratitude to their creator—those aspects of human beings' usufructuary mediality which draw attention to their subordinate position are most important. For Hale, both in the passages discussed above and in those which will be discussed later in this chapter, it was more important to emphasize the limits that God placed on their relative position of power over the rest of creation. For both writers, as for the other writers cited in this chapter, usufruct offered a clear and consistent conceptual framework on which to build their discussion of humanity's mediality and its ramifications.

Displacement established the structure of possession underlying the usufructuary ethos; *mediality* has described the nature of the relationships that usufructuary displacement instantiated, and established mediality's pervasiveness in seventeenth- and eighteenth-century writing. The next and final core concept of the usufructuary ethos, *accountability*, will begin to lay out the specific moral consequences that seventeenth- and eighteenth-century writers extrapolated from humankind's usufructuary status, and will further explore the way that legal metaphors functioned as moral arguments in religious writing.

ACCOUNTABILITY AND *THE GREAT AUDIT*

Limits, obligations, and boundaries to human use of and power over the nonhuman world characterize usufructuary mediality. But in order for boundaries and obligations to be meaningful, people must be held *accountable* for the uses they make of their power and their possessions, and for whether or not those uses fulfill the terms under which they were granted. Under the usufructuary ethos, human beings are accountable *to* the outright owner (usually, but not exclusively, God) *for* the future health of the earth. Notice that the *to* and the *for* split usufructuary accountability into two distinct but related threads. In the first, the usufructuary is accountable for staying within the limits of acceptable use set by the proprietor; stepping outside those limits violates the proprietor's ultimate authority. In the second, the usufructuary is accountable to *future users* for preserving the earth in its original state of health. Usufructuaries are thus accountable *to* both the proprietor and future users *for* the land and its future health.

In late seventeenth-century English writing, human beings' accountability for their use of the nonhuman world was most often expressed in the legal-social terms that were also central to expressions of usufructuary mediality. Hale's *The Great Audit* provides an especially powerful example of the precise way human beings' medial position as usufructuaries made them accountable to God for his creation. Like "usufructuary" and "steward," the term "audit" carried legal, fiduciary, and religious connotations. Per the *Oxford English Dictionary*, "audit" could refer to a judicial hearing, a "periodical settlement of accounts between landlord and tenants," or a "searching examination or solemn rendering of accounts, *esp.* The Day of Judgment."[40] Hale invokes all three senses simultaneously, figuring as a part of the spiritual "rendering of accounts" an examination by the landlord God into how his tenants and stewards—humans—have used and cared for his creatures.

Hale begins by describing the relationship among God, humans, and nonhuman creatures as a *trust:* "I received and used thy creatures as committed to me under a Trust, and as a Steward and Accomptent for them; and therefore I was always careful to use them according to those Limits, and in order for those Ends, for which thou didst commit them to me."[41] He then describes two categories of "Limits" and "Ends" that humans will be held accountable for under the terms of God's trust. The first category has to do with *what* the steward did with God's creatures, and with their motivations for their actions. The human steward is charged, says Hale, with using creatures with "Temperance and Moderation," for the "Support of the Exigencies of [his] Nature and Condition," and not to "Luxury and Excess, to make Provision for [his] Lusts." Though Hale, like any contemporary moralist, stresses that the sinfulness of "Luxury and Excess . . . Lusts . . . vain Glory or Ostentation" is a function of gluttony, he attributes the wrongness of such misuse, and the consequences of it, not merely to the fact that gluttony is a sin, but specifically to the fact that it represents a violation of man's duties as a steward and usufructuary. Whenever eating or drinking, Hale says, "I checked myself, . . . still remembered I had thy Creatures under an Accompt; and was ever careful to avoid excess or Intemperance, because every excessive Cup and Meal was in Danger to leave me somewhat Insuper and Arrear to my Lord."[42] Thus using creatures to excess is not simply a sin of personal gluttony; it violates the terms under which God granted humans their limited dominion, by (in this case, literally) eating into God's resources.

Both "insuper" and "arrear" are the language of indebtedness, but indebtedness in the sense of a debt incurred and still to be paid, or of a failure to fulfill a duty.[43] "Insuper" was a term used in auditing to mark outstanding accounts;[44] to be "Insuper" for "every excessive Cup and Meal" thus implies the wrongful consumption of that which belongs by right to God, a violation of the divine law of limited use. "Arrear" had similar connotations of monetary or material debt, but it also referred to falling "behind in the discharge of *duties*," making the taking of "excessive Cup and Meal" a double violation of God's trust: the act of taking more than you have a right to, and the failure to act with the "moderation and temperance" that is your duty as "Steward and Accomptant." Hale's use of both "insuper" and "arrear" signals the density of his sense of human accountability. They are accountable to God as his trustees in the sense that at the termination of the trust (their death), they will be called to account for their use of nonhuman creatures. The "accounting" for their use will take place as an account (both a story and a counting) of what they used and how much, and whether they overstepped their bounds. Finally, they are "accountable" in the sense of being responsible for observing the limits and discharging the duties that were assigned to them. To be God's steward, accomptant, viceroy, usufructuary—to be given the right and ability to use his creation—is to be accountable to him in all these senses.

The second category of "Limits" and "Ends" Hale attaches to the "Trust" of nonhuman creatures concerns what constitutes just use of the nonhuman creatures that God "hast put under [human] Power and Disposal." Humans must use the creatures "With Mercy and Compassion" for the "Powers of Life and Sense" which they possess.[45] Those "Powers" ought not be taken from them merely in service of human appetites. Crucially, Hale frames the issue of exercising "dominion" with "mercy and compassion" explicitly in terms of the legitimacy of political power. Failure to exercise mercy and compassion toward other living beings transforms rightful usufructuary lordship to tyranny. Hale cautions those wishing to be good stewards to remember that though God "has given us a Dominion over thy Creatures, yet it is under a Law of Justice, Prudence, and Moderation; otherwise we should become Tyrants, not Lords, over thy Creatures."[46] Besides not inflicting unnecessary or unnecessarily painful deaths on nonhumans for human subsistence, Hale also states that the "Law of Justice" makes it a human duty to care for nonhuman creatures during their lives: to "deny domestical Creatures their convenient Food; to exact that Labour from them that they are not able to perform; to use Extremity or Cruelty towards them; is a Breach of that Trust under which the Dominion of the Creatures was committed to us, and a Breach of that Justice that is due from Men to them ... to be merciful to [their] Beasts." To use or treat other creatures in a way that violates the terms under which God granted human usufructuary power is therefore "a Tyranny inconsistent with the Trust and Stewardship that thou [God] has committed" to human beings.[47]

The distinction Hale draws between tyranny and stewardship turns on the issue of accountability. A tyrant acts without deference either to the authority that granted him his trust or to the duties he owes to those over whom he has power. His exercise of power is unjust because it fails to acknowledge that it is fundamentally limited and accountable. A lord, in contrast, understands that his power is granted temporarily and under specific terms of stewardship and care, and that he is accountable to God for those that have been entrusted to him, and for faithfully fulfilling his obligations. In other words, man turns from "lord" of nature to "tyrant" over nature when he forgets that both his possession and his power are limited and medial. Hale believes the power granted to humans as the stewards and usufructuaries of the world to be by its very nature subject to a law whose primary purpose is to ensure that justice and happiness is, overall, extended to each creature. That is, after all, the rationale that licenses human vice-regency in Hale's *Primitive Origination*: that they maintain God's ideal balance among the competing needs of various creatures for the optimal happiness of all. Only by justly fulfilling the duties laid out for them by God can humans legitimately claim "dominion" over the nonhuman world. Hence, the contractual relationship Hale imagines existing among God, humans, and nonhumans makes humans ethically responsible for the subjective well-being of His creatures, as well as for maintaining the integrity and health of the natural world as a whole.

Once again, Hale's position is not unique. As Keith Thomas has observed, that "view of man's relationship to animals would have a long life" and a wide array of supporters from the seventeenth century onward,[48] and "justice" and "tyranny" are the very terms most often used to describe proper versus improper exercise of human dominion over nonhuman creatures in the seventeenth and eighteenth centuries. Like Hale, Thomas Tryon contrasts proper dominion over animals with tyranny in the opening paragraph of his 1684 husbandry manual *The Country-Man's Companion*: "The Righteous Man (saith the inspired Prophet) is Merciful to his Beast: . . . 'Tis generally said, and very truly, That Man is the Vice-Roy of the Creation, and to him is given Dominion over the Beasts of the Earth; but this Rule is not absolute or tyrannical, but qualified so as it may most conduce . . . to the helping, aiding and assisting those Beasts, to the obtaining all the Advantages their Natures are by the great, bountiful and always beneficent Creator made capable of."[49] Tryon not only equates "absolute" dominion with tyranny but also defines man's proper dominion as "qualified" (as befits the usufructuary ethos), and furthermore qualified specifically by the obligation to ensure the well-being of other creatures. Similar to Hale's view of "vice-regency" as distinct from tyranny, Tryon's "Righteous Man" is an active steward, responsible for guaranteeing that creation operates as God intended: to the cumulative happiness of all creatures, achieved when every part of creation obtains "all the Advantages their Natures are . . . capable of." As in Hale, tyranny consists not simply of cruelty or mistreatment, but of behaving as though human power is "absolute" rather than usufructuary, overstepping the boundaries that define just use of power in order to satisfy human appetites and desires. Tryon describes the same basic usufructuary structure of power that Hale did: accountable to God for the care of his creation.

Usufructuary Accountability and English Law

By invoking legal and political concepts to reinforce the divine laws governing human use of the nonhuman world, Hale and Tryon suggest that the structure of rights and accountability among God, humans, and the nonhuman world is analogous to that among tenants, landlords, and the authority of the Crown and courts to enforce tenants' compliance with law and equity. In fact, Hale ties the obligation to treat laboring animals with kindness directly to human obligations of justice and humanity to other humans: "I have ever thought that there was a certain Degree of Justice due from Man to the Creatures, as from Man to Man; and that an excessive, immoderate, unreasonable Use of the Creature's Labour, is an Injustice for which he must account."[50] More than a mere turn of phrase, such language draws parallels between human and divine justice, and between social and environmental hierarchies. Just as it is punishable under human law to damage property entrusted to you and belonging to another, so is human destruction of God's property punishable by God. In all such cases, you will be held to account for exercising your authority with justice by the entity who has authority over you.

The logic behind that argument is crucial to understand. The sort of natural jurisprudence Hale describes derives its form and its efficacy from taking the analogies likening divine, social, and environmental relationships literally. It's not just that the relationships among beings in the natural hierarchy are *like* those in the social hierarchy—they are the *same*. In *The Gentleman's Calling*, Allestree makes that exact claim, once again relying on the legal vocabulary of trusts and stewards: "God has placed *Man* in the World, not as a *Proprietary*, but a *Steward*; he hath put many excellent things into his possession, but these in trust" to God.[51] Like Hale, Allestree couches his explanation of this particular set of human duties in legal terms. He compares the relationships among humans, God, and the nonhuman environment to a "person having, as in our Law-forms is usual, covenanted to stand seised of the Estate," and by so doing, having agreed to put it only to "the proper uses, to which it is to be limited." Those limitations center on using the land "moderately": a good steward of God "falsifies no part of his trust, nor abuses his stewardship, this being, as it were, the allowed Fees of the place, a Pension allotted to him by the bounty of his Lord."[52] For humans to engross more power or goods to themselves than God intended is tantamount to a breach of the duty owed to God as his usufructuary stewards.

Allestree's legal language is considerably less precise than Hale's; unsurprising, given that Allestree was a churchman, and Hale one of the leading legal minds of his day. *The Gentleman's Calling* uses "fees," "estates," and "trusts" more or less interchangeably, but for Allestree, what all those types of legal ownership had in common was that they were medial and accountable to the authority that granted them. Human beings' power as stewards, as envisioned by writers like Allestree and Hale, was limited and accountable to the source of their (revocable) power. More importantly, Allestree's conflation of different types of English law under that assumption, and the way he likens them to human beings' positions vis-à-vis God and creation, reflect the pervasiveness of the usufructuary ethos across multiple levels of discourse and thought. For Allestree, true ownership is always displaced, the human position always medial, and human beings always accountable for using their power and their property correctly, on environmental, social, and legal levels.

It was no coincidence that philosophical and religious texts were stacked with English legal metaphors alongside references to usufruct. As David Worster has noted, the fundamental assumption underlying property rights in England in the early modern period was that of usufructuary ownership.[53] It was technically still true in the eighteenth century that every lord in England was subject to the authority of the Crown, which was, according to feudal law, the true, ultimate owner of the land. John Cowell's *Institutes of the Laws of England* (first Latin edition 1605), one of the major sources for William Blackstone's seminal *Commentaries on the Laws of England* (1766),[54] points out, "Since there are none in *England* besides the Soveraigne power, who hath a plenary and absolute dominion over immoveables, it is not hard to discern who they are, that are the Possessors of estates, as to the profits, the estates not being fully theirs, which we shall tearm *usufructuarii*."[55] Cowell lays out here the usufructuary nature of land ownership in England and the key role displacement of ownership from tenancy played in it: because only the "Soveraigne power" of the Crown can have "absolute dominion," all English land "owners" are technically usufructuaries, their "estates not being fully theirs" in the sense of having total freedom to do and dispose with as they like. Blackstone echoed that truism at the beginning of his discussion of property law in the *Commentaries*, noting that it is a "received, and now undeniable, principle in law, that all lands in England are holden mediately or immediately of the king.... A subject therefore hath only the usufruct, and not the absolute property of the soil."[56] Combined with Blackstone's invocation of the displacement principle as the philosophical basis for human law, English land law is doubly usufructuary in his analysis: held first of God, and then of the king.

Of course, as Blackstone is quick to point out, fee simple had by his time become, for legal purposes, almost the same as outright ownership, and therefore the king's role as ultimate proprietor of all English property was a technicality with little or no practical legal relevance.[57] Nevertheless, the notion that English landholding could be described through the concept of usufruct retained currency even in legal circles. Eighteenth-century lawyers before and after Blackstone continued to refer to the similarity between fees and usufructs when parsing the jurisprudential foundations of English land law, always returning to the notion that there existed in English law a responsibility to other parties for the stewardship of the land, and that that responsibility had its root in the usufructuary split between owner and tenant. Despite being an imperfect analogy for the kinds of property English law actually created—it did not quite fit "fee simple," the most basic and complete form of English landownership, nor the estate for life or term of years, to which it was most frequently compared, nor the more modern "trust" and "use"—usufruct was invoked again and again as a way to better understand the implications of those property forms. The reason has less to do with practical matters of law than with jurisprudence and with concerns about the ethics both of the land laws themselves and of landownership in general that were taking shape after the Restoration. In legal texts, as in devotional texts—as well as in the natural law texts both borrowed from—usufruct provided a way to clarify the relationships among people and property in terms of their rights of use and their duties to one another, in the present and across time.

For example, John Cowell's 1651 *Institutes of the Laws of England* used usufruct to explain the difference between fee simple and other forms of property rights. Cowell explains that in fee simple, the "possessor hath a perpetuity," whereas "he onely is an usufructuary in an Estate with us [in

English law], who hath Lands or Tenements for Term of years, or at the will of another, or who hath Lands by way of pledge or security which we call Mortgage, or . . . lastly, he who hath the Lands of an Heir in the right of Guardianship, untill he comes to full age."[58] In all of these cases, the operative distinction that Cowell draws attention to is between the fee simple, in which possession and use are in the same hands, and other types of property rights such as the estate for life or term of years, the use, or guardianship, in which case the party using the land is different from and accountable to the party who holds the permanent right to the land. Cowell continues to refer to the holder of a life estate, the beneficiary of a use, and any of these other types of owner collectively as "the Usufructuary" through the rest of the discussion, a term that conveys that what is most important about these forms of property from a jurisprudential standpoint is the displacement of ownership away from use, and the existence of someone to whom the usufructuary is accountable for the use of the land. In cases such as this, the comparison to usufruct provides clarification of the principles and structures of ownership underlying that particular bit of common law. The estate for life and the use are unique English property forms, and the concept of "usufruct" is meant to help Cowell's reader understand precisely the relationships among lessor, lessee, and the land which that form of fee entails, and to ground the principles that differentiate the lessee from the holder of the fee simple in the codes of Roman civil law.[59]

Of course, in order to be held accountable for the use of something, there have to be consequences for its misuse. In English law, this took the form of the holder of a life estate or use being prosecuted for *waste*—the legal term for misuse, mismanagement, or damage to property through either neglect or action, by a tenant who is responsible for its maintenance to its heir. The concept of waste was also tied directly to usufruct in English jurisprudence. *The Landlord's Law*, a popular handbook of English property law, explains that "all Damnification, Detriment, or Imparation of a Thing out of its natural and proper Use is Waste, because the *Tenant only hath the present Usufruct*, and the Reversioner a Right to the same and the like Use on his Decease, or any other Determination of his Interest."[60] This description uses usufruct to explain the legal principles behind waste: it is a neglect or misuse that violates the right of future holders to inherit the land as it was when the tenant took possession. "Damnification, Detriment, [and] Imparition" are waste "*because*" the tenant is by definition a usufructuary. Usufruct here provides the jurisprudential basis which both explains and justifies the English rule of waste.

The comparison went the other way as well, with texts on civil law noting usufruct's similarity to English common law life estates, a pattern of conceptual cross-referencing that suggests the depth of the cultural association in early modern England between the two historically unconnected legal concepts. In a supplemental note to his section on usufruct in his 1734 *New Pandect of Roman Civil Law*, John Ayliffe argued that having "Lands and Tenements for Life" in English law "is almost the same as an Usufructuary in the *Roman* Law. . . . he who has a Freehold, is subject to an Action of Waste, and may forfeit the same by a Mal-User thereof; which could not be, if he was the absolute Proprietor of the Land, and had the free Disposal thereof."[61] That is to say, a life estate and a usufruct share the fundamental premise that possession of land is limited by the existence of other parties to whom the holder bears a responsibility for the integrity of the estate; thus, argues Ayliffe, the fundamental legal premises of rights and duties are the same in the usufruct and the life estate.[62] And, to return to the broader point of this chapter, both *The Landlord's Law* and Ayliffe show that just as the usufructuary is in a medial position between the outright owner *to* whom he is responsible and the land *for* which he is responsible, the English tenant of a life estate or term of years was a sort of middleman, through whom property passed from past to future; and he was directly accountable for making sure that it made that transition intact.

Accountability to future generations was emphasized even further in the eighteenth-century development of the so-called equitable waste. Life estates and uses were both situations that created what is called an "equitable" estate, distinct from the legal estate. While a legal estate was, literally, an estate (e.g., fee simple or fee-tail) as constituted, recognized, and enforced by the common law as established through precedent and statute, an equitable estate was recognized and enforced by the Court of Chancery, which was empowered to make rulings on possession and waste not merely on the letter of the law, but on the basis of fairness and justice to the rights of those with current and future interests in the land. Owing to the legal technicalities of common law tenures and the quirks of the common law legal process, there were many cases in which the misuse or destruction of property in which another had a future interest was, technically, legal. Yet its legality did not make it fair or right. This disparity led to Lord Chancellor Hardwicke's development of the doctrine of equitable waste, which placed equitable injunctions on legal actions of waste committed by life tenants, under the premise that such waste violated the rights of those who would later inherit the estate to receive that estate intact and undamaged. Equitable waste recognized that those without legal rights to sue for the destruction of property nevertheless may have ethical rights, and made it possible to appeal for relief in chancery court on the basis that otherwise legal action could prove deleterious to the long-term preservation of the land. Essentially, equitable waste acknowledged the existence of ethical obligations inherent to life tenancy that existed above and beyond those set out in common law statute.[63]

At the heart of eighteenth-century discussions of English land law, therefore, rests an assumption that the landlord occupies the same medial position with regard to power and land that the traditional usufructuary does: he is a steward accountable *to* others with rightful claims to the land *for* their responsible, sustainable use of it. Those with rightful claims included both the present holder of the fee (as in legal waste), as well as those who will rely the property in the future (as in equitable waste). The comparisons to usufruct in legal writing thus helped to elucidate and reaffirm the legal and ethical stakes of the relationships among land and people that legal situations like the life estate instantiated, explaining and justifying the principles behind those common legal instruments. At the same time, the comparison brought English law into the web of beliefs and associations that made up the usufructuary ethos. Like the devotional writings discussed above, usufruct operated in legal writing as a metaphor that drew out the mediality and accountability that governed English relationships to both their property and their authority. References to usufruct pointed to the fact that there were ethical and moral implications to the temporary and limited nature of power and possession, and that particular legal types of possession were analogous to those assigned to human beings in the ontological hierarchy of being.

For all their differences in approach and subject, whether lawyers or divines (or both), Anglicans or Dissenters, all the authors discussed so far converge on the same basic usufructuary principles. What you have is "yours" only in a partial and temporary sense. That means that there are moral limitations restricting how and why you may justly use what you (temporarily) have. It also means that you must manage your land with care for the benefit of those others who depend on it, and of those to come. The writers in this chapter return again and again to the same basic assumptions of displacement, mediality, and accountability in their discussions of the rights and duties associated with the human possession of power and property, invoking *usufruct* and the vocabularies associated with it—trust, steward, accountant, lord—to express and enforce those rights and duties. Their common vocabulary reflects the interconnections among religion, law, philosophy, and morality in early modern England, a time and place where "lord" and "tenant" could apply equally to human beings in society and to God and his creatures.

This, finally, is what makes the usufructuary ethos an ethos, rather than merely a set of metaphors, or an unconnected series of moral admonishments. The concept of mediality and the value of accountability are fundamental to the ways that English people understood their identities and their relationships with all other living beings in the present and across generations. The core belief that power and possession were displaced, limited, and temporary formed the basis for a set of ethics that sought to govern not only relationships among human beings but relationships among humans and nonhuman beings, both divine and terrestrial, for all time. It is, inherently and unavoidably, a hierarchical worldview, since it is premised on the presumption that humans have been rightly gifted with greater power and responsibility than other earthly creatures (as well as that some human beings were rightly granted more power and property than others). The usufructuary ethos enshrines that inequality of power and privilege in its very core; but it also contains within it a unique ethics of use, a set of rules about what kinds of uses power and material resources might justly be subjected to and which justifications for use are valid. It provided its explicators with a way to articulate to their audiences the moral duties incumbent upon them based on their particular positions in the world. "Usufruct" lent conceptual depth to their arguments, as well as the authority of a tradition of natural law that placed usufructuary displacement at the heart of human morality. It was, as we shall see, a paradigm of ethical thought with a long and influential history through the first half of the eighteenth century.

LOCKE'S USUFRUCTUARY PROPERTY VERSUS THE USUFRUCTUARY ETHOS: TENSIONS EMERGE

Locke appeared at the beginning of this chapter among the thinkers who rooted their theories of natural law in the usufructuary displacement of ownership. Locke's theory of property in the *Second Treatise on Government* begins with the same assumption of God's permanent ownership as do all the other writers in this chapter: recall Locke's insistence that "Men being all the Workmanship of one Omnipotent, and infinitely wise Maker; . . . they are his Property, whose Workmanship they are, made to last during his, not one anothers Pleasure."[64] The link between "Property" and "Workmanship" becomes the foundation of Locke's theory that because "every Man has a *Property* in his own *Person*," the "*Work of his Hands, we may say, [is] properly his*," and therefore anything he "has mixed his *Labour* with" becomes his property.[65] Given Locke's insistence in both the *Second Treatise* and *Essays on the Laws of Nature* that God's ownership of his "Workmanship" is permanent, we can therefore conclude that Locke's laboring man occupies a classic usufructuary medial position, both owned and owner. What's more, the *Second Treatise* famously includes two so-called provisos that hold the property owner to account for his use of his natural resources. The first is the *waste proviso*, which requires that an individual leave "enough, and as good . . . in common for others" rather than appropriating all available resources to himself. The second is the *spoilage proviso*, which states that an individual can rightfully claim only as much as he can "use . . . before it spoils."[66]

The waste and spoilage provisos in the *Second Treatise* underscore the deep influence that the usufructuary ethos had on Locke's seminal theory of property, society, and power. The spoilage proviso in particular invokes usufructuary implications for the responsibility of Lockean landowners to the public and future generations, and resonates with the language used in contemporary discussions of usufruct: "The same Law of Nature, that does by this means give us Property, does also *bound* that Property, too. . . . But how far has [God] given [the earth] us? *To enjoy*. As much as any one can make use of to any advantage of life before it spoils, so much he may by his labour fix a property in: whatever is beyond this, is more than his share, and belongs to others. Nothing was made by God for Man to spoil or destroy."[67] The proviso states that the law of nature upon which the human right to property depends establishes specific "bounds" from its inception. There are echoes of Sir Matthew Hale and other contemporary devotional writers' discussions of man's usufructuary status in Locke's insistence that those bounds that are based in the fact that God, to whom all of creation ultimately and permanently belongs, has only given the earth to humans to a limited extent. It was given "to enjoy," but *not* to "spoil or destroy."

Intriguingly, the same language of "enjoying" without "spoiling" subsequently crops up in English texts on civil law usufruct later in the
eighteenth century, suggesting that some readers perceived a conceptual
connection between Locke's proviso and the theological-legal concept of
usufruct. A 1729 translation of Pufendorf's *De Jure Naturae et Gentium*
by Basil Kennett, for instance, defines usufruct as "the right to use and
enjoy the things of another without impairing the substance," echoing
Locke's word choice and adumbrating the same guidelines as Lockean
property: yours to enjoy, as long as your enjoyment does not damage it.[68]
The most direct linguistic parallel between Lockean property and the
definition of usufruct, however, comes from a 1722 translation of Jean
Domat's *Les Lois Civiles dans leur Ordre Naturel* by William Strahan.
Originally published in France in 1689 on commission for Louis XIV,
Domat's systematization of the Roman civil law quickly became the authoritative text on civil law, serving as the basis in the later eighteenth
century for legal reforms including the Code Napoleon and the German
civil code.[69] Strahan's 1722 edition translated Domat's definition of usufruct as the "Right to use and *enjoy* a Thing which is not our own, preserving it whole and entire, without *spoiling*, or diminishing it."[70] Like Locke,
Strahan uses the verbs "to enjoy" and "to spoil" to define, respectively, the
rights of the landholder and the parameters limiting those rights. That
direct linguistic parallel indicates not only that Locke's theory of property in the *Second Treatise* is structurally usufructuary but, more specifically, that that structural similarity was recognized by writers and readers
at the time. The fact that Strahan's translation was published thirty-three
years after the first edition of the *Second Treatise* connects Locke and
his *Second Treatise* to the evolving vocabulary of the usufructuary ethos,
providing another set of terms to describe the moral limits of the human
usufructuary right to use creation: to enjoy, without spoiling.

Scholars have recognized for some time both the usufructuary nature of Locke's property and its environmentalist potential. Environmental philosophers including Kristin Shrader-Frechette and Rebecca
Judge have argued that the provisos constitute a "sustainability con-
straint"[71] on the right to property which applies to future generations
as well as to those who are currently alive, meaning that the Lockean
duty not to "spoil or destroy" preserves property rights even of those who
are not yet born.[72] James Tully,[73] Clark Wolf,[74] and, most recently, Zev
Trachtenberg have explicitly pointed out the usufructuary aspects of
Locke's theory of property. For Trachtenberg, the usufructuary qualities

of Locke's theory of property are what make it a potentially powerful environmental paradigm. He argues that "Locke's theory doesn't grant ownership of the productivity of nature itself," which remains God's, but rather that "God grants human being *usufructory* right to Earth—rights to use the products of nature, but not absolute ownership of the underlying natural systems that ultimately provide for humanity's survival by generating those products."[75] Because God granted usufructuary rights to the products of the earth for subsistence to *all* people equally throughout time, Trachtenberg concludes that "Locke's theory suggests . . . that unsustainable practices are violations of natural law. The obligation to the future implicit in the usufructory character of property reinforces the notion that . . . [e]arlier generations have no greater claim on the goods of the Earth than later generations; hence they have an obligation to support their own survival in ways that don't foreclose the prospects of their descendants."[76] Trachtenberg's model draws out key elements that Locke's theory of property has in common with the usufructuary worldview: absolute ownership is displaced from human hands to God's; human possession is limited and medial, passing through the present generation to the future; and the present generation of users is accountable both to God the owner and to future generations for using the land as God intended—for the benefit of all, present and future. Trachtenberg reads Locke in isolation, but his intuitions about the usufructuary implications of the *Second Treatise* are reinforced when placed in its full historical context. More broadly, the fact that Locke's theory bears similarities to the usufructuary ethos evinced by so many other popular authors of his time suggests, as I pointed out earlier, that the usufructuary ethos contributed to the way Locke was interpreted by his contemporaries in ways not yet recognized.

Yet in many important respects, Locke's theory diverges from the usufructuary ethos as the writers in this chapter understood it. One major reason for the divergence connects back to the tension between Pufendorf's and Locke's understandings of human moral obligation in their use of nonhuman beings. Though both rejected the notion of human "moral community" with nonhumans, Pufendorf insisted that humans owed a moral duty to God not to use his creatures "idly or wantonly," whereas Locke asserted that what humans did with nonhumans only mattered insofar as it affected other humans' rights. The disparity between Pufendorf and Locke on this issue points to an underlying tension developing over the seventeenth and eighteenth centuries between theories that conceived of property as conferring a set of *duties* on the possessor versus those that conceived of property as conferring *rights* to the possessor.[77] In Pufendorf's more traditional view, "'property has its duties'; an adage which seems to capture the most humane elements of the medieval world."[78] Pufendorf's emphasis on duties corresponds loosely with the usufructuary ethos, particularly in the way the notion of property having "its duties" defines property by the relationships it creates—between God and human, between human and human, *and* between human and nonhuman. In this view, property exists in the context of community.

In contrast, Locke follows Hugo Grotius in articulating a view of "a less interdependent social world: property has its *limits*."[79] Though Locke's provisos have their basis in God's original gift of the earth to all in common, they are primarily meant to protect each individual human being's property rights, not necessarily to steward God's original order. Rather than placing its possessor in relationship to others to whom he owes duties, Locke sees property as an extension of the individual. Locke's provisos are meant first and foremost to protect that individual's rights. The waste proviso, which requires leaving "enough, and as good" for others, is intended to protect other individuals' right to appropriate; if you appropriate too many resources for your own property, you invade other individuals' right to appropriate resources as their own property. Likewise, while the spoilage proviso protects against a transgression against God by destruction of his property, it is primarily meant to prevent transgressions against other people, because hoarding and wasting resources they could have used prevents them from asserting their own rights to appropriate and benefit from natural resources. For Locke, the plants or animals or lands that were wasted are irrelevant; he repeatedly insists that nonhuman beings have no claims of any kind, moral or material, on human beings. Humans' only obligation is a negative one: not to violate other humans' rights. Even Trachtenberg concedes this problem: "[I]t would be a mistake to claim [Locke] as an early environmentalist. . . . Locke did not understand the pervasiveness, complexity, scale, and . . . range of organic and inorganic processes." His lack of "ecological awareness," Trachtenberg concludes, leads him to underestimate the importance of nonhumans to the land's productivity and to misconstrue the degree to which individual choices about land use can have cascading effects on others' property.[80] That issue, however, is the result of Locke's total dismissal of nonhuman beings as having any significance to either

God or man as anything other than raw material. Even Locke's concern for future generations is ultimately about protecting the property rights of individual humans.

Still, the strong conceptual and linguistic connections to the usufructuary ethos in Locke's *Second Treatise* make it impossible to sever his theory from its usufructuary roots. That inseparability points up an important complication for the usufructuary ethos. Namely, the bedrock assumption of usufructuary displacement did not necessarily lead all early modern writers and thinkers to the same moral conclusions. Locke develops similar concepts around boundaries and limits to possession and use, but both his premises and his conclusions differ from those of writers like Allestree or Hale, resulting in theories of possession and use that are in tension with, and in some ways even antithetical to, what this chapter has defined as the usufructuary ethos. Usufructuary displacement could underpin the emerging liberal ethos of individualism and personal accumulation in the *Second Treatise* at the very same time that other writers used it to underpin the usufructuary ethos of cross-species mutuality and obligation. In that sense, Locke's *Second Treatise* could be read as an early liberal instantiation of the usufructuary ethos, a continuation of a still-common set of socio-environmental values that are in the process of being reshaped to fit a newer, emerging set of political and economic conditions and values. That fact demonstrates how deeply embedded usufruct was in English thought. But it also throws into relief the growing pressure being exerted on the usufructuary ethos by the philosophical, political, and economic developments of the late seventeenth century.

Money, America, and Usufructuary Ethos

The *Second Treatise* points toward a second challenge to the usufructuary ethos: the rapidly changing economics of late seventeenth-century England. As the *Second Treatise* continues, Locke finds loopholes for his own provisos in the technology of money and the development of settler colonialism. As the previous section illustrated, the spoilage proviso is the key to establishing the possibility for there to be ethical limits to personal accumulation in Locke's theory of property. For example, Locke wrote, "He that *gathered* a Hundred Bushels of Acorns or Apples, had thereby a *Property* in them. . . . He was only to look that he used them before they spoiled; else he took more than his share, and robb'd others. And

indeed was a foolish thing, as well as dishonest, to hoard up more than
he could make use of."[81] The most important part of this passage for the
present discussion is the way spoilage functions as a naturally occurring
limit to accumulation. Locke argues that no one has a right to squan-
der something that could be of use to someone else; spoilage, the loss
of the potential of that item to support the life of another being, provides
the point at which a thing has officially been squandered.

But when money was invented, its durability rendered the spoilage
proviso obsolete. Locke continues:

> If he would give his Nuts for a piece of Metal, pleased with its co-
> lour; or exchange his Sheep for Shells, or Wool for a sparkling Pebble
> or Diamond, and keep those by him all his Life, he invaded not the
> Right of others, he might heap up as much of these durable things as
> he pleased; the *exceeding of the bounds of his* just *Property* not lying in
> the largeness of his Possession, but the perishing of any thing uselesly
> in it.
>
> And thus *came in the use of Money*, some lasting thing that Men
> might keep without spoiling, and that by mutual consent Men would
> take in exchange for the truly useful, but perishable Supports of Life.
>
> And as different degrees of Industry were apt to give Men Posses-
> sions in different Proportions, so this *Invention of Money* gave them
> the opportunity to continue to enlarge them.[82]

Locke's argument demonstrates the way money enables value to
shift from subsistence to greater and greater levels of abstraction and
ephemerality—first metal coins, then eventually paper—and in so doing
introduces a wrinkle in the usufructuary ethos. Through the invention of
money, Locke argues, "Men have agreed to disproportionate and unequal
Possession of the Earth," famously establishing the basis for the development of economic individualism.[83] With that shift comes an important
shift in the moral basis of possession: whereas the use-it-or-lose-it logic
of the spoilage proviso makes sharing obligatory, the invalidation of the
spoilage proviso by the introduction of money removes that obligation.
"What reason could any one have . . . to enlarge his Possessions beyond
the use of his Family" in a place where the only goods are perishable,
Locke asks?[84] None. But when money is introduced, and possessions are
no longer subject to spoilage, the way is paved for the moral accumulation of wealth.

Of course, Locke's theory of property features two provisos, not just one, and the "storage of wealth in the form of money" makes it easier for people not to leave "enough, and as good" for others.[85] As Herman Lebovics points out, the "social and economic setting in which Locke wrote the *Second Treatise*—a conjunction of intensified development of the nation's land, now overwhelmingly in the hands of private owners, with a large landless population of poorly-paid laborers . . .—renders paradoxical a theory of property" that purports to base ownership on labor. The way Locke gets around this paradox, as Lebovics and Jimmy Casas Klausen have shown, was through an "America" that is empty and available for appropriation: "there was land, quite enough and very good, in the New World."[86] Throughout his discussion of money and individual accumulation, Locke returns to the idea of "America," which in the *Second Treatise* was a place where people had infinite supplies of land and "no hopes of Commerce with other Parts of the World, to draw Money to [them] by the Sale of" goods.[87] For Locke, the lack of money or commerce explains indigenous people's supposed lack of private property, and their lack of private property and subsequent failure to maximize the land's productivity justify English appropriation. Thus "Locke managed with the same argument both to justify the dispossession of the ancestral lands of [indigenous Americans] and the ongoing enclosure of the commons once set aside by custom for the use of the peasant of the English countryside":[88] the enclosure of land in England and the accumulation of wealth by a few does not violate natural law because there is "enough, and as good" in America; and there is "enough, and as good" available in America because the indigenous people there do not engage in the kind of labor that removes the land from the commons. Through first money and then America, Locke manages to neutralize his own provisos, and with them, usufructuary limitations on his version of property.

The way that Locke invokes usufructuary values only to devote considerable effort and space to explaining how money and settler colonialism render them obsolete points toward the status of the usufructuary ethos in late seventeenth- and early eighteenth-century England, as well as toward some potential reasons for the preoccupation with it among certain writers. The fact that Locke felt the need to couch his theory in usufructuary terms indicates their centrality to the expectations and assumptions of his readers. Human beings' usufructuary possession of the earth was the basis for English concepts of property, duty, and rights, a basis Locke shared. Even the provisos show that Locke accepted that usufructuary displacement implied certain ethical ramifications. On the other hand, the fact that the *Second Treatise* culminates in a theory of property and political society that sets aside the usufructuary ethos on the basis of the conjunctions among money, enclosure, and settler colonialism is no coincidence, coming, as it did, from a man who was not only one of the architects of the Financial Revolution but also a financial investor in numerous colonial ventures in the Americas. The *Second Treatise* places colonial mercantilism in tension with usufructuary values, and eventually articulates a way for the former to supersede the latter without invalidating it entirely. Locke does not debunk the usufructuary ethos, he merely explains why money and America make it irrelevant.

Locke was not the only writer to perceive a tension between the usufructuary ethos and the economic and colonial developments of the period. In fact, one major factor in the rise of the usufructuary ethos to such a level of cultural prominence at this particular historical moment was the anxiety that that tension produced. There is no need to convince people to abide by moral guidelines implicit in a particular set of values if you trust that they already are. But it *is* necessary if, in spite of the persistence of the cultural authority of those values, other social, economic, or material conditions exert a countervailing pressure against them. Writers in the late decades of the seventeenth century and early decades of the eighteenth were preoccupied with the question of what those who had greater "gifts" of property and power ought and ought not to do with them. The possibility of the wealthy classes accumulating greater and greater property to themselves, and of developing rhetoric undermining the traditional socio-environmental communal duties governing its use, posed a significant and growing threat to the usufructuary ethos. Locke's *Second Treatise* represents the most famous (and perhaps most influential) version of that rhetoric, and it illustrates that that rhetoric existed in the context of a response to the usufructuary ethos. But while Locke sees these developments as a boon to society, others saw them as a threat.

Anxiety about the problem of misuse of riches, whether through avaricious accumulation or prodigal waste, was a common theme well into the eighteenth century in all kinds of writing. This discourse had roots in traditional exhortations of moral behavior for the rich and powerful, but it took on a particular intensity and shape in response to the financial and commercial transformations that continued to unfold over the course of the first decades of the eighteenth century. As the avenues and opportunities for profit widened, and novel financial instruments such as credit and transatlantic joint stock companies rose to prominence, it became necessary to make finer distinctions between what kinds of "gain" were acceptable or unacceptable. It also became necessary to articulate the usufructuary duties landowners ought to be fulfilling, and to delineate the pitfalls of failing to do so, in more stringent and detailed ways. Always at the heart of the texts examined in this part of the chapter is an awareness of the problem of power: belief in and adherence to usufructuary duties can be urged, chided, commended—but it cannot be compelled.[89]

An example from a 1722 book of meditations on Proverbs 31 by a writer named Oswald Dykes offers a direct look at the ways early eighteenth-century English writers interpreted the relationship between new economic developments and the usufructuary ethos. A former amanuensis to the outspoken Jacobite Sir Robert L'Estrange, Dykes was deeply suspicious of the social and moral repercussions of the economic innovations of his age.[90] He is reluctant to condemn financial profit outright, however, and the distinctions he draws among methods of getting and using property will begin to illustrate the complicated ways that defenders of the usufructuary ethos sought to reconcile usufructuary values with their social and economic reality. In *The Royal Marriage*, Dykes uses a commentary on Proverbs 31:26 as a point of departure for a wide-ranging discussion of the importance of good stewardship and diligent labor, and of the evils of usury, debt, and the emerging new money economy.[91] As a piece of that discussion, Dykes invokes usufructuary obligations to both posterity and public. "It is the Father's great Duty to preserve the Patrimony of his Predecessors," Dykes writes, "as by a lineal Descent from Ancestors to Heirs, intirely without any Waste or Impov'rishment; Mortgage, Incumbrance, or Alienation: for the Benefit of his Wife and Children; as a faithful Usufructuary, or an accountable Guardian of their future Good." Dykes's definition of a "faithful Usufructuary" as an "accountable Guardian of . . . future Good" who also "preserve[s] the Patrimony of his Predecessors" links together the usufructuary values of mediality and accountability with an emphasis on the connection of past with future via family and present stewardship. Likewise, Dykes also emphasizes the importance of "Improving" the land with an eye to "publick as well as private Profit." He decisively disallows avaricious accumulation in the name *only* of family: a "Man's Acquisitions are highly discommendable, and become selfish or blame-worthy, in Acting only for his own mercenary Ends, without any farther Views of his Neighbour's

Benefit."[92] Neither spending nor saving is correct in itself; rather, Dykes locates the proper use of riches in the balancing of personal and public needs, in the present and future.

But not all forms of "improvement" of the estate are equal in Dykes's eyes. Though a High Tory, Dykes does not draw that line strictly on the basis of land versus money, but rather on his own adjudication of "honest" versus "dishonest gain." That adjudication—the notion that some kinds of "gain" were morally defensible, others morally pernicious—and its connection to the South Seas Bubble—gesture toward the ways that the intertwined forces of colonialism and financial revolution began to reshape the usufructuary ethos in the early eighteenth century.[93] "Improvement" is accomplished, Dykes explains, "by Care, Travail, and good Husbandry," either through "natural" means like tillage or the "artificial" means of "Buying or Selling for honest Gain." Dykes privileges "good Husbandry" as the "most commendable Study, fruitful Imployment, and pleasant Business" of all, besides "Learning, Morality and Religion."[94] Still, commerce and trade, he recognizes, are legitimate ways to "improve" the estate by increasing its monetary profitability—so long as it is done for "*honest* Gain," by which he means gain that will provide benefits for posterity and the public.

Dishonest gain is not only selfish and avaricious but, more importantly, ephemeral. Dykes repeatedly opposes bogeymen of the economic revolution like the "Stock-Jobbers of Beggary in a Common-Wealth" and the "*Cannibals* of Mankind about *Change-Alley*" against "*Agriculture* and *Gardening* . . . the happiest Diversions of humane Being, Business, or Society."[95] For Dykes, the most important difference between land and financial instruments like stocks is that the former will persist reliably through time, whereas the latter are fickle and ephemeral, and have the power to transform present and future stability into nothingness. Thus, when she finds herself with surplus capital, the ideal wife

buys Land; that which is solid, and not *chimerical* or *aerial Bubbles* in City, Town or Country. She neither builds nor buys any groundless Castles in the Air. *Terra Firma* is what she wants to purchase. She loves to lay-out her Money upon substantial Certainties, never subject to any cross Accidents, or the fickle Contingencies of Fortune. . . . Nothing can hurt Land, generally speaking, but Want of good Manuring and proper Tillage, or oppressive Mortgages. . . . She knows they breed a Canker in a flourishing Estate; which either ruines the present Possessor, or wrongs the lawful Inheritor. . . . But, of all Things, she desires to have no dealings in Paper or Wood, in Tallies or Chalk-Scores. . . . She delights neither in the grand or petty Lotteries of the Times, even in Hope of getting the great *Benefit-Ticket*. . . . She does not understand the prodigious Rise and lofty Down-fall of *Stocks* at Direction by the Will and Pleasure of a few ingenious, fraudulent, fleecing Projectors.[96]

The Royal Marriage appeared two years after the South Seas Bubble, and Dykes clearly has that incident in mind as he hammers away at the reliability and solidity of landed wealth over monetary wealth. Land, family, and long-term environmental and economic stewardship take on greater urgency in the wake of what Dykes clearly considers to be the cataclysmic repercussions of moving away from the more (theoretically) traditional model of familial, social, political, and environmental stability and well-being inhering in the possession and proper stewardship of the land itself by ideal landlords. Implicitly, Dykes recognizes that the sort of socio-political-environmental stability he strives for depends on the gentry classes of England focusing on stewarding their ancestral lands for the future. The threat of *"chimerical* or *aerial Bubbles"* like the South Seas Company inheres in the potential for financial instruments like joint-stock companies—or mortgages—to transform land into money, and for money to turn into nothing. That is the key similarity between stocks and mortgages: mortgages trade land for ready money, to satisfy a current need or want, at the expense of keeping that property secure for the future.

As J. G. A. Pocock and others have argued at length, that view was a common one among writers of the late seventeenth and early eighteenth centuries, who furthermore were suspicious of the complicated relationships among Crown, Parliament, the Bank of England, and other new creditors who turned a profit on the government's debt. In the eyes of those like Dykes, commerce was corrupting and the disinterested devotion to the public good was associated with land.[97] What I am suggesting is that another piece of Dykes's extreme anxiety about and almost hysterical vilification of the "Stock-Jobbers of Beggary" and the *"Cannibals* of . . . *Change-Alley"* derives from his realization that the financial transformations happening around him have the power to destabilize the usufructuary ethics of possession upon which an entire shared socio-environmental understanding of the world relied. Dykes sees the (relatively) short-term profits and seeming ephemerality of the new economy as a threat to the traditional landed economy, but beyond that, he sees it as a threat to England's socio-environmental order as well, present and especially future. Financial instruments like mortgages and speculative investments meant to yield personal monetary profit mitigate against being a "faithful Usufructuary" and "accountable Guardian of . . . future good" in Dykes's eyes because they render fungible and ephemeral the embodied basis of the socio-environmental community: the shared land on which all depend. The thick entanglements in this discourse among lords, fathers, land, patriarchy, finance, and environment are typical of the usufructuary ethos of the landlord. But in Dykes's defensiveness, we can begin to see the cracks forming.

Indeed, once we know what to look for, works that express anxiety about the potential for those with property and power to flout their usufructuary duties appear everywhere. The devotional works this chapter has focused on clearly described and strongly endorsed the usufructuary ethos, but other writers grappled with the messy collision of precept with reality. William Wycherley laid out the core of the problem with accumulation that flouts usufructuary duty in the title of an obscure poem in his 1704 *Miscellany Poems*: "*To a Covetous, Rich, Proud Man, who us'd to say*, He was only Frugal, to make what he had Last; *and said*, He was only his own Steward, but that he might be sure not be [*sic*] Cheated." The poem describes a wealthy landowner who is intent upon an extreme frugality that, Wycherley emphasizes, cheats himself, his dependents, and God of the true, collective benefits of wealth. Using the multivalent language of "accounts" that we encountered earlier, Wycherley exhorts the rich man to "cast up your Account," both monetary and spiritual, and "see, to what your Stores amount." If he wishes truly to make the most of his gifts, Wycherley argues, "and multiply [his] Store, / Lend not the Rich, but give now to the Poor."[98] The failure to do so is the rich man's greatest sin, and will represent the largest deficit on his spiritual ledger:

> Heav'n; whose Bounty too you but abuse,
> Which of its Loans expects so Pious Use;
> . . . .
> . . . but once cast up true your Account;
> You'll find, that much is wanting to the Poor,
> All to the Rich, who runs with Heav'n o'th'Score,
> Repays it less, as his Debt's to it more.[99]

The foundational error of the covetous man Wycherley addresses is his failure to recall his usufructuary status, that his property is merely a "Loan" from God, granted to be used, not hoarded against some kind of theoretical future personal dearth, when a need for its use exists in the present. Wycherley does not propose a solution, nor is his poem didactic in the sense of attempting to teach a specific moral lesson. Rather, his poem suggests that the rich man's individual error exposes a more troubling problem: the rich man has the power to make a disastrously "covetous" choice, and he can exploit usufructuary rhetoric to justify himself. What's more, the anonymity of Wycherley's portrait, combined with his use of the indefinite article in the title—he is *a* covetous, rich, proud man, not *the* covetous, rich, proud man—implies that he is describing a type rather than an individual. The problem his poem diagnoses, of the landowning class's failure to operate with a proper sense of their medial role in the social and ontological hierarchies, is endemic, not particular, and as such it has deep and long-lasting implications for English society, politics, and landscape.

CONCLUSION

Insofar as it upholds an ontological hierarchy that assigns humans more power than nonhumans, and insofar as it is an ethic of power and use that applies equally to political, social, and religious relationships among humans as well as among humans *and* nonhumans, the usufructuary ethos as it has been described in this chapter may not seem explicitly or particularly "environmental." The chapters that follow will delve into the environmental implications of the usufructuary ethos as they manifested in the poetry of the long eighteenth century. The environmental ethics of that era emerged out of, and are expressive of, the beliefs that shape that culture's definitions of justice and rightful power in specific and direct ways. That may seem like an obvious argument—of course the environmental ethics of a certain time and place will be a part of the larger culture of that time and place—but in the case of the usufructuary ethos of the seventeenth and eighteenth centuries, what we will find is that writers applied the *exact same* structure to the ethical relationships among humans and nonhumans as they did to relationships among humans, enfolding the environment into broader sociopolitical questions of justice, power, duty, and obligation. And at the very heart of the structure of the usufructuary ethos are ideas recognizable as environmental values for twenty-first century environmentalism: that the world does not belong to us to do with whatever we wish; that there are moral constraints on the ways we may use and treat nonhuman beings; that we are responsible for making sure the world remains sustainable for the benefit of future generations.

At the same time, as the second half of this chapter has shown, the usufructuary ethos was as much a reaction to anxieties about real and potential abuses of power, property, and nature as it was a set of positive values. The impulse to express usufructuary values arose from a sense that those values were under threat from competing ways of defining an ethical relationship between humans and the natural world. The following chapters turn to the poetry of the late seventeenth and early eighteenth century, because it is the poets who wrestled with the question of how to square the usufructuary values they still held with the social, economic, and environmental changes they were experiencing—changes they often only ambivalently opposed. The devotional and philosophical writers studied in this chapter set the terms by which the morality of the use of property and power were to be judged, not only for the purposes of this book, but for much of the English population of the time. The purpose of their writing was to articulate clearly (to paraphrase Allestree's famous title) the duty of man. The writers in the following chapters hold that lens up to the world.

2

Trees, Posterity, and the Socio-Environmental Landlord

I N *The Whole Duty of Man* (1657), his ubiquitous mid-seventeenth-century book of devotional essays, Richard Allestree wrote that it is "folly . . . to be proud of the Goods of Fortune; by them I mean Wealth and Honour, and the like. . . . We have them but as Stewards, to lay out for our Master's use, and therefore should rather think how to make our accounts, than pride ourselves in our receipts."[1] Allestree uses the language of stewardship and accounts that the previous chapter established as central to the usufructuary ethos, and it reflects the ethos's three core tenets: *displacement* ("the Goods of Fortune . . . Wealth and Honour, and the like" are fundamentally the property of God); *mediality* (humans are "Stewards," not owners, and have only the use of those gifts from God, placing them in between divine authority and the "Goods" they have usufructuary rights to use); and *accountability* (humans will have to "account" for their use of the "Goods of Fortune" to God). More importantly, this passage exemplifies the way the usufructuary ethos functioned as an ethics of *use*. Allestree adumbrates a set of parameters for the proper use of the nonhuman world structured around the assumption that humankind's usufructuary possession is predicated on the duty to use what they have been entrusted with for the ends which their "Master" intended. That which humans will be held accountable for is how well they fulfill their duties as stewards of their portion of God's creation. By extension, the greater the portion of creation a person has in his usufructuary possession, the greater his duties.

No group in England possessed greater "Goods of Fortune" than the landed upper classes. As Allestree later wrote in *The Gentleman's Calling,* "those *Advantages*, by which [gentlemen] are severed and discriminated from the vulgar, . . . consequently by being peculiar to them, devolve on them an obligation of a *distinct Duty*."[2] That duty, Allestree goes on to explain, involves maintaining his estate for the present and future support of his dependents, human and nonhuman, rather than exploiting its potential for short-term personal enrichment or enjoyment. In keeping with his admonitions in *Whole Duty*, Allestree insists that the landed gentleman must prioritize what will be recorded in the divine "account" of his uses and duties, rather than the profits or benefits he personally stands to reap. Thus, the ways that devotional texts such as Allestree's align human beings' role in the environmental hierarchy both linguistically and conceptually with ideas of sociopolitical lordship is not coincidental; it is part of a worldview that sees the natural and social orders as analogical to one another. For Allestree, as for the other writers that will be discussed in this chapter, landlordship carried with it specific social and environmental obligations that were of great moral importance.

The cultural, social, political, and environmental importance of the usufructuary ethos coalesced in the late seventeenth and early eighteenth century in the figure of the landlord, the person who had received the greatest gifts from God, and with them the greatest responsibilities. Though its precise expression differs from writer to writer, a core idea echoes through all the texts in this chapter: that the natural world itself embodies the intergenerational and social communities to and for which the landlord is accountable, and that therefore the ways that the landlord uses his land carry political, moral, and environmental significance. Throughout the period, the usufructuary landlord was considered accountable to both posterity and the public for ensuring the ongoing existence of an environment that would sustain denizens of all classes and species. The figure of the usufructuary landlord is related to J. G. A. Pocock's "ideal of the citizen, virtuous in his devotion to the public good and in his engagement in relations of equality and ruling-and-being-ruled," which Pocock argues emerged around the same time, but the usufructuary landlord's environmental commitments were more explicit, and more literal.[3] From the portrayals of the landlord and the use of riches examined in this chapter emerges a unique socio-environmental ethic that draws on the usufructuary values of mediality and accountability to drive home the moral imperative for landlords to steward the nonhuman environment responsibly for the public and posterity. Yet it is a socio-environmental ethic that is, by definition, exclusive to the landowning class of men. These texts express a strongly conservationist mindset, prioritizing the long-term well-being of living things both human and non- above considerations of personal profit or enjoyment, but they place the power and right of ensuring that well-being in the hands of landlords, and landlords only.

The first section of this chapter lays out the specific role and responsibilities the landlord had in the hierarchy of being according to popular devotional writing of the late seventeenth century. Devotional writers insisted that all beings in the social and ontological hierarchies had particular responsibilities they were obligated to fulfill. In the case of the English landlord, his responsibilities were to ensure the well-being (1) of all the other living things that depend upon his estate (what is often called the "public") and (2) of future generations (his "posterity"). The second and third sections of this chapter read texts by John Evelyn and Anne Finch on landlords and trees through the lens of the usufructuary landlord. In both Evelyn's *Silva* and Finch's "Upon My Lord Winchilsea's Converting the Mount in His Garden to a Terras," trees take on additional importance as organisms whose unique biological traits create both symbolic and ecologically literal connections among other beings and across generations—*if* landlords protect and steward them properly. Most critical attention to these texts has focused on the military, mercantile, and political implications of England's post-Restoration timber shortage, or on teasing out the ways Evelyn and Finch deployed trees to symbolize the triumph of the Tory aristocracy and the Stuart line. That focus, I argue, has tended to overemphasize the role of trees as props in the political dramas happening around them at the expense of the importance their status as living beings had to the authors who wrote about them. By reapproaching these texts in the context of the usufructuary ethos, this chapter also reexamines the connection between the political ideology of late seventeenth-century English writers and the nascent conservationism that has been located in Evelyn and Finch by some environmental and eighteenth-century scholars.

The final section will turn to John Philips's 1708 georgic poem *Cyder*. Recent work in eighteenth-century studies has centered the georgic mode as the period's most important contribution to environmental literature, but thus far *Cyder* has received little of that attention.[4] *Cyder* offers an early glimpse into how the classical georgic mode was adapted to reflect and support the eighteenth-century English socio-environmental context laid out thus far in this study. Philips's *Cyder* was the first adaptation of the georgic to merge the genre's classical authority and ethos of forward-looking care with the usufructuary ethos's concerns with environmental duties to public and posterity and the ethical question of the use of riches. The final section of this chapter will both situate *Cyder* within the usufructuary ethos and show that the poem's political and environmental points of view are inextricable from one another.

In arguing that these authors' inescapably hierarchical and patriarchal views carry important and overlooked environmental implications, I do not dismiss the crucial insights of scholars including E. P. Thompson, Raymond Williams, John Barrell, and James Turner into the ways that depictions of landscape and landlords in early modern writing worked to naturalize the exploitation of labor and the consolidation of property and power in the hands of the few.[5] It is certainly true that all the writers in this chapter believed in, benefited from, and perpetuated the notion of a natural hierarchy that explained and licensed social hierarchy. What I am arguing is that that hierarchical sociopolitical worldview had its roots in the hierarchical logic of the usufructuary ethos, and that that ethos produced a specific form of historical environmentalism, in which the living estate takes on simultaneously moral, social, political, *and* genuinely environmental significance for the human being responsible for them. This chapter outlines the contours of that usufructuary socio-environmentalism, including both its concern for future sustainability and its inherent classism, beginning with its foundation in devotional writing.

However, as the previous chapter suggested, the turn to the usufructuary ethos by the authors in that chapter was always predicated on the perception that those around them were straying from it. Evelyn's *Silva* inveighs against clear-cutting trees to plant grain for export, a growing practice in late seventeenth-century England. In that practice, Evelyn sees the lure of short-term profit (tied, not coincidentally, to colonial expansion, as much of that grain was shipped across the Atlantic) competing with the long-term, cross-generational duties of the socioeconomic landlord. Finch's poem likewise was written on the occasion of the loss of a landscape to the myopia of a lord. And in *Cyder*, Philips transforms a georgic topos, that of the book 4's industrious bees, into a cautionary tale of the avaricious and short-sighted wasp whose "lawless Love of Gain" compromises the stability of the community.[6] Thus, even as these texts offer a glimpse of an idealized socio-environmental community and landlord, they are shadowed by the knowledge that their ideal is vulnerable to the whims of individual landlords, who may or may not be willing to sacrifice their own profits or pleasures for the sake of public and posterity. Usufructuary socio-environmentalism, therefore, is from its inception shaped by the forces that threaten to undermine it.

Seventeenth-Century Devotional Writing and Ontological Hierarchy

In the seventeenth century, writers perceived a direct analogy between roles and duties in the natural world and those in the social world. As Courtney Weiss Smith has recently shown, writers from Boyle to Newton to Bolingbroke and Pope believed that close attention to the particularities of the natural world yielded not only valuable empirical information but theological and political insights as well.[7] That epistemological habit of deriving theological and social truths from natural observations was no less common among clergymen. As the last chapter demonstrated, popular devotional works like Thomas Adams's sermon "The Barren Tree" used examples of the earth's obligation to yield sustenance to the creatures that depend upon it in order to remind readers that humans beings also were obliged to yield up their fruits to God in exactly the same way and for the same reasons.[8]

Adams also applied the analogy of the fruits to the realm of human social relations. Just as each being has an obligation to fulfill their role in the natural hierarchy, each person has a social obligation to fulfill the duties incumbent upon people of their rank in order to support other members of the community. Each member of society must yield up their particular "fruites," according to Adams: of the "Magistrate ... the *fruites* of Justice"; of the "Minister ... the *fruits* of knowledge"; and more generally, of the "private man ... the *fruit* of his calling: to bee idle, is to bee barren of good; and to bee barren of good is to bee pregnant of all evill." The last of these is a lesson aimed particularly at the rich, and Adams turns to the particular fruits expected of wealthy Christians: "Let us all produce the *fruits* of Charity: ... The *fruite* of Christianitie is Mercie; when the rich, like full eares of Corne, humble themselves to the poore earth in Charitie. Feed him, that feeds you: give him part of your Temporalls, from whom you expect Eternalls. ... Our mercie to others, is the *Fruite* of Gods mercy to us." Nothing, Adams insists, "is created for it selfe, but so placed by the most wise providence, that it may conferre something to the publique good."[9] This applies as much to social relationships as interspecies relationships. The important point here is that *all* types of relationships, for writers like Adams, operate according to the same basic usufructuary structure. The landlord is like an ear of corn, and vice versa, in that they each have something that will support and sustain others, and therefore an obligation to share it. Moreover, the greater the power an individual has, the greater their degree of obligation to share. To quote Luke 12:48, the implicit message is: "For unto whomsoever much is given, of him shall be much required: and to whom men have committed much, of him they will ask the more."[10]

As his title suggests, Allestree's *The Gentleman's Calling* focuses on the particular social and ontological "callings" of those who have been "given much"—namely, landowning men. The fifth section of the treatise, "On the Advantage of Wealth," focuses on the "proper uses" to which the possession of "riches" (specifically land) "is to be limited." Allestree begins by reaffirming the lord's usufructuary right to benefit from his land through another analogy to the earth itself: "like the Earth which though she conveys the springs through her veins, yet is allowed to suck in so much, as may give her a competent refreshment," gentlemen are permitted to "retain" a part of the benefits of their wealth for themselves. But, Allestree continues, employing the quasi-legal vocabulary of the usufructuary ethos, only he that benefits from his property "moderately, and with a thankful reflection on that liberal Providence which thus *gives him all things richly to enjoy*, 1 Tim. 6.7, falsifies no part of his trust, nor abuses his stewardship; this being, as it were, the allowed Fees of his Place, a Pension allotted him by the bounty of his Lord."[11] To use riches "moderately," Allestree implies, is to stick to one's duty to manage his resources responsibly, and to avoid both overuse and miserliness as a matter of justice. Those duties discharged, the gentleman may claim whatever is left without danger of encroaching on God's limits.

What Allestree subsequently goes on to describe is a set of duties that literally frame the use and stewardship of land and property as an issue of moral accountability to both God and other living beings. The connections that have been elsewhere implicit between humans as usufructuary lords of the earth and as lords in the sociopolitical hierarchy of England are here direct: To be a member of the land-owning gentry is to be a "steward" and "usufructuary," holding the earth in "trust" not only for God but for the "publick" and for "posterity"—that is, for those (human and nonhuman) beings below him in the social and ontological hierarchies, and for future generations of beings. The "publick" portion of this duty,

for Allestree, consists of a fairly standard case for the social importance
of those of a higher class in supporting the needy, enforced by an appeal
to the usufructuary principles of displacement and mediality. The rich
are "meer Receivers" of wealth from God for the purpose of "the care of
the Poor," and the rich man's possession is "never so absolute, as to ex-
clude the power of revocation" by God for failure to discharge his fiduciary duties.[12] Thus, Allestree asserts, "Every rich Man is, as I said before,
God's Steward, and particularly instructed to provide for the indigent
parts of his Family"; if he fails to do so, he "invades" the rights of those
who needed the resources he wasted, and "abuses" God's "trust."[13] Allestree's insistence on humankind's temporary, incomplete possession and
its duty to use those riches to help others points back to the hierarchical understanding of the universe in which each individual has a specific
role to fulfill to maintain the harmonious functioning of the social and
natural worlds. His use of legal language to buttress that claim, however,
once again indicates that these claims are not about behavioral norms,
but about ethics. To fail to fulfill one's duty is an act of injustice.

The wealthy's role as caretakers for the needy is a well-established justification for the unequal distribution of riches. What's less well known,
particularly among ecocritics, are the specific ramifications this socio-
environmental hierarchy had for the definition of "use" and "usefulness"
in the long eighteenth century. In "On Mutual Subjection," a sermon
written sometime in the 1720s or 1730s and published posthumously in
1744, Jonathan Swift echoed both the point Allestree made in the previous paragraph and the language of the trust that Allestree and other
seventeenth-century devotional writers used to describe the relationship
among God, humans, and material "riches."[14] Swift states that the "Advantages of Life, such as Riches, Honour, Power, and the like" were "de-
posited with" the rich "merely as a Trust . . . to be employed for the Use
of his Brethren."[15] Like Allestree, Swift emphasizes the rich man's role as
God's trustee, and the role of "his Brethren"—those dependent on
him—as the beneficiaries of that trust. However, Swift also draws out
an important component of the issue of use and usefulness that is more
oblique in Allestree and Adams. He defines "usefulness" explicitly as a
quality that *every* being must have, including wealthy human beings, and
links the fulfillment of ontological usefulness to moral character:

As God hath contrived all the Works of Nature to be useful, and in
some Manner a Support to each other, by which the whole Frame of

the World, under his Providence, is preserved and kept up; so, among Mankind, our particular Stations are appointed to each of us by God Almighty, wherein we are obliged to act, as far as our Power reacheth, towards the Good of the whole Community: And he who doth not perform that Part assigned him, towards advancing the Benefit of the Whole, in Proportion to his Opportunities and Abilities, is not only an useless, but a very mischievous Member of the Publick.[16]

To be "useful" means to "act . . . towards the Good of the whole Community"—what Adams, Allestree, and Swift also term the "public." Failure to fulfill your duty—to be "useless"—is morally wrong, because that failure poses a threat to the community's well-being and cohesion. Swift's is a slightly different version of Adams's argument that human beings must be like the earth or cows: given a gift, you must make use of it, and make yourself useful, not just for your own benefit, but for the benefit of society and of the rest of creation.

This is the understanding of use and hierarchy that has generally been occluded by twentieth- and twenty-first-century assumptions about the concepts. The typical narrative has held that seventeenth- and eighteenth-century Europeans saw the nonhuman world as an inert resource that existed solely for their anthropocentric benefit, and that they believed that they were above other nonhuman organisms in virtually every sense, owing to their superior rank in the chain of being.[17] All of that is true, to some extent. The piece of the picture that has been missing, and that this section has been laying out the contours of, is the co-existing belief that human beings were a fundamental *part of* that hierarchical universe, not separate from it.[18] Humans shared with every other organism a role in a hierarchy that they were not separate from or at the top of, and because of that role they too had an obligation to use their gifts correctly, and to be of use themselves.

The mistake of much previous scholarship on environmental attitudes of the period has been to presume that most people believed hierarchy and use primarily to confer privilege on human beings, rather than obligation. On the contrary, central to the conceptions of hierarchy and use at work in these seventeenth- and eighteenth-century authors is *mutuality*. Hierarchy is a shared condition, and use is about sharing, care, and support. So consistent was this belief in the human/lord's role as the cornerstone of the social and environmental order throughout the eighteenth century that Johnson used a quote from Swift's "On Mutual Subjection" in his *Dictionary* entry on "Steward" to emphasize human beings' role as the trustees and stewards of God, tasked with looking after his creatures: "what can be a greater Honour than to be chosen one of the Stewards and Dispensers of God's Bounty to Mankind? What can give a generous Spirit more Complacency than to consider, that great Numbers owe to him, under God, their Subsistence, and the good Conduct of their Lives?"[19] Swift's sermon remained in print in various forms for the rest of the century, which, with Johnson's use of it in the *Dictionary*, points to a persistent belief in the landlord figure's central role in guaranteeing the well-being of the "public" writ large. Because of his place in the environmental and social hierarchy, the landlord had a duty to make himself and his gifts "useful" to the survival and well-being of other creatures.

Furthermore, the landlord's duty of justice and care was an *intergenerational* one. His material obligations existed not just among the currently living but across time, through the shared dependence of multiple generations on his usufructuary property. At the beginning of the "Advantages of Wealth" section of *The Gentleman's Calling*, Allestree writes that he "must mention one part of Duty, as fundamental to all the rest. And that is the well-husbanding of [his land]; not in a figurative, but real sence, the having such a provident care of those goods and possessions, wherewith God hath blest a man, as may secure them from . . . Consumption . . . in *Justice to a Man's posterity*."[20] Failing to steward property with due care—whether through neglect or greedy overuse—is thus a double violation, according to Allestree, both of the terms of God's trust and of the rights of future generations. Echoing Sir Matthew Hale's *The Great Audit and the Good Steward*, Allestree turns to the notion of the account to drive home the point that failure to maintain the earth with an eye to its long-term sustainability is tantamount to intergenerational injustice: "And what tolerable account can such a Parent give to his beggar'd Offspring? . . . Nay, what account can he give to God, from whom he received it, in order to several ends, if he thus at once defeat them all? 'Twas a Command to the *Jews*, that they *should not cut down Fruit trees, though it were for so necessary an use as to advance a Siege, Deut.* 20.19. But this is the hewing down that stock, from which so much good Fruit should spring, and that without any pretence either of necessity or reason."[21] Allestree's use of "account" language still underscores the landlord's fiduciary role, as Hale's did, but in this case he combines the trustees to whom he is accountable: both God the proprietor *and* future generations of users. The landlord must, in other words, see himself as a link in a temporal as well as a social and a species-based chain of being. Allestree thus merges human beings' medial role in the chain of being—passing the means of survival from one being to the next—with their concomitant role as a link in a *temporal* chain that connects past and future through the present. The fact that Allestree invokes Deuteronomy to substantiate his argument for the gentleman's obligation to posterity alongside his usual legal metaphors highlights the way the usufructuary ethos united social, legal, religious, and natural duties under a single ethical obligation: to guarantee the future by preserving what one received from the past.

Allestree was a leader in the post-Restoration Church of England and a lifelong royalist, but the idea of humans' accountability to the public and to posterity for the use and conservation of the earth crops up in works by writers across the political and religious spectrum. Around the same time that Allestree published *The Gentleman's Calling,* John Flavel, a popular Presbyterian minister who had been ejected from his parish in 1662 as a result of the Act of Uniformity, published *Husbandry Spiritualiz'd.* The book, which (like *Gentleman's Calling* and "On Mutual Subjection") remained in print throughout the eighteenth century, contained a chapter, "Upon the Husbandman's Care for Posterity," that used the "husbandmen's" literal obligation to "not only labour to supply their own Necessities while living, but to lay up something for their posterity when they are gone" as an analogical argument for the duty to inculcate proper religious devotion in children (thus the conceit of the book, "Husbandry *Spiritualiz'd*").[22] The chapter ends with a poem that frames the current generation's spiritual duty to future ones in terms of its literal duties of land management:

> A Publick Spirit scorns to plant no Root,
> But such, from which himself may gather fruit.
> For thus he reasons, if I reap the gains
> Of my laborious Predecessors pains,
> How *equal* is it, that Posterity
> Should reap the Fruits of present industry?
> Should every Age but serve its turn, and take
> No thought for future times? It soon would make
> A Bankrupt World, and so entail a Curse
> From Age to Age, as it grows worse and worse.[23]

Flavel's poem once again makes the repercussions of present-day use an issue of justice. Unlike Allestree and Swift, Flavel relies not on God's intentions to make his case but rather on the practical and ethical duties incumbent on the present generation, and he does so in a way that ties past, present, and future together. His insistence that it is "equal"— here meaning fair or just, as in the courts of "equity"—for "Posterity" to "reap the Fruits of present industry" on the basis that the speaker reaped the "gains" of his "laborious Predecessors pains" points to an underlying assumption of intergenerational justice rooted in the use and care of the created world all generations live in and with. It is just, Flavel implies, that each generation should work to reproduce the benefits they enjoyed from the work of the past. To fail to do so, he goes on, would have disastrous practical repercussions: a "Bankrupt World," and the entail—the inheritance—of a "Curse," rather than a blessing. Thus, like Allestree, Flavel envisions the present generation of humans as the medium through which the world passes, with the duty and the power to ensure the stability and well-being of the future generation. Flavel's appeal to the "Publick Spirit" that "scorns to plant no Root" further points to his understanding that these duties extended beyond the immediate family, forming the basis for a fully functional society. The concept of "public" itself is intergenerational. In this prominent thread of belief, therefore, human beings were the usufructuaries not just of God, but of the public and of posterity, all at once.

The Socio-Environmental Ethic of the Landlord in Evelyn's *Silva*

In both examples of intergenerational accountability offered in the previous section, the authors referred to trees to drive their point home. Allestree appealed to an Old Testament commandment against cutting down fruit trees for any reason, including military purposes. Flavel's reference was more metaphorical, playing on "root" and "fruit" to emphasize the connection between a landowner's choices and his community's collective well-being, as well as to denounce the impulse to focus on short-term gain over long-term benefit. Both examples draw on trees' long lifespans and capacity to provide sustenance and support to multiple groups and generations in order to exemplify the intergenerational connections that the reader is being exhorted to honor and conserve. Trees, in these examples, embody the interconnections among landlord, public, and posterity that undergird the usufructuary socio-environmental ethic of the landlord.

That both Allestree and Flavel turn to trees to make this point is not
a coincidence. These devotional writers were drawing on an important
way that trees signified in late seventeenth- and early eighteenth-century
writing, as both symbols and embodiments of public and intergenerational interdependence. As the anthropologist Laura Rival has observed,
"Euroamerican consciousness and sense of identity is deeply marked by
the fact that [trees] outlive human beings. . . . [Trees] encompass human
history as living links not only to the past but also to the future" thanks
to their extensive lifespans and human communities' deep historical
dependence on them.[24] This section will examine the significance of
trees as emblems of the intergenerational socio-environmental relationships among the landlord, his dependents, his ancestors, and posterity.
It will first examine the period's most famous and influential work on
trees, Evelyn's *Silva*, and then an especially tree-centric poem by Anne
Finch, "Upon My Lord Winchilsea's Converting the Mount in His Garden to a Terras." For Evelyn and Finch, trees are the living, literal locus
of the entangled lives of all beings across species, time, and rank—tree
and human, past and future, lord and tenant. Finally, in both Evelyn and
Finch the landlord's exclusive right and duty to plant and protect trees
reveals the way conservationism and classism were inextricably entangled
in the usufructuary ethos. The landlord has an inherent duty to sustain
the land for future generations, but he is also, by definition, the only one
capable of carrying out that task. In Evelyn and Finch, we see examples of
the ways that the merged social and environmental hierarchies described
by Adams, Allestree, and Swift shaped representations of the relationships among humans, their environments, and class in other genres of
writing, tying the responsibility for and act of conservation explicitly to
landowners.

The original impetus for John Evelyn's *Sylva* (as the title was originally
spelled) is well-known.[25] The Royal Society received a request from the
commissioner of the navy to do a study of the state of England's forests
and timber reserves in September 1662, a task Evelyn, along with other
members of the Royal Society, took up.[26] England's woods had been rav-
aged by decades of cutting. There was growing concern about the nation's
ability to support the needs of its growing naval fleet in the coming years.
Despite the politically motivated habit among royalists of blaming the
crisis on the depredations of the Interregnum, forest clearing had been
a growing problem under Charles I and royalist landowners before the
Civil War as well. Evelyn recognized this, though he was of course far

more sympathetic to the "'gallant and loyal gentry' who had added to this waste to 'preserve the poor remainder of their Fortunes'" than to the Interregnum's "late prodigious spoilers ... unhappy Usurpers and injurious Sequestrators."[27] His exhortations to the landowners of England in *Silva* to preserve England's extant forests and plant new ones, therefore, were aimed at convincing the landowning class as a whole to return to the usufructuary ideals of stewardship for public and posterity that Evelyn frames as traditional.

Before ever mentioning the navy's needs, Evelyn describes tree planting as an act of transgenerational care, and he frames the landlord as the medium through whom responsibility for the care of England's forests passes. In this sense, *Silva* functions as a case study in what the usufructuary ethos looked like in the late seventeenth century in a nondevotional context, when its purview is narrowed to the question of the role and duties of one particular rank of human being, the landowning lord, in relation to other social and natural ranks of beings. In the preliminary section entitled "To the Reader" Evelyn's first move is to adumbrate a continuous tradition of tree planting by landowners beginning in antiquity and continuing to the present, and then to place responsibility for the continuation of that tradition in the hands of his landowning reader. The preservation and repair of England's forests, Evelyn writes, "is what all Persons who are *Owners* of *Land* may contribute to ... who are touch'd with that laudable *Ambition* of imitating their illustrious *Ancestors*, and of worthily serving their *Generation*."[28] Evelyn suggests that planting trees and caring for them enacts the living connection between the past and the future by consciously repeating the actions of "Ancestors," which ensures a sustainable world for their "Generation." For Evelyn, the tree planter is a generational medium. Through him passes both the tree-planting tradition and the guarantee of a tree-filled environment from one generation to the next. *Silva* thus portrays the landowner as a version of the usufructuary medium described by Sir Matthew Hale in the *Primitive Origination*.[29] In fact, Evelyn may have had that exact comparison in the back of his mind when he was writing *Silva*: he owned a copy of Hale's *Primitive Origination*, and he marked the passage that refers to humans as God's usufructuaries and viceroys.[30]

Evelyn develops the importance of living nonhuman beings with far more depth than Hale had, however. Trees are the simultaneously literal and symbolic connectors of generations across time. *Silva* depicts both the tree planter and the actual trees that he plants as temporal media, as an anecdote Evelyn later recounts shows: "When *Ulysses*, after a ten-years Absence, was return'd from *Troy*, and coming home, found his aged *Father* in the Field planting of *Trees*, He ask'd him, why (being now so far advanc'd in Years) he would put himself to the Fatigue and Labour of Planting, *that* which he was never likely to enjoy the Fruits of? The good old Man (taking him for a Stranger) gently reply'd; *I plant* (says he) *against my Son* Ulysses *comes home.*"[31] Ulysses' initial confusion about why someone would bother to plant trees he personally would never live to benefit from points up the displacement principle at the heart of Evelyn's argument: the landowner should think of his property not as something that is *his*, but as the source of support of all beings present and future, in particular his posterity. Ulysses's father is thinking on a scale that stretches beyond a single human's lifetime. That is precisely why trees recur in both *Silva* and other English writers as an emblem of transgenerational sustainability. Ulysses's father plants *trees* against his son's return, as opposed to grain or flowers, because trees' long life-spans make them a biological touchpoint that can literally connect beings across time, even beyond the human lifespan. The same organism—the same tree—exists for decades or even centuries; children and grand-children and even great-grandchildren of the people alive when it was planted can eat the same tree's fruit or sit under its branches. The Ulysses anecdote merges the literal, biological traits of trees with their symbolism of generational connection, making the planting of *trees*, and the particularities of trees as organisms, central to their significance. The ecological web and the generational web merge. Ulysses's father's trees are simultaneously literal and symbolic, because it is their actual mode of existence in the world that makes them emblematic of the continuity of human generations.

The Ulysses story is a part of *Silva's* sustained effort to reframe its readers' perceptions of the unique role that living trees have played in improving and preserving social life and cohesion across time, and the unique role that the landlord has played as the planter and preserver of those trees. Throughout *Silva*, Evelyn seeds in anecdotes that reinforce the importance of trees not only to the landowner's own children but also to the welfare of the community as a whole. He cites authorities from Cicero to the Duke of Luxembourg to argue for the "publick Calamity" of destroying woods, and commends the practice of an *"Italian Nobleman"* who commanded *"an Hundred Thousand"* trees be planted on the occasion of his daughter's birth. Planting trees, Evelyn goes on, is a

"most *Charitable* Work . . . for the benefit of the *Poor,* upon *Commons,*
and other waste Grounds, and such places where they would thrive . . .
for the use of the Indigent."[32] It is a far better use of the rich man's resources, Evelyn notes, than horses and dogs. In the chapter on walnuts
in the 1706 edition of *Silva,* Evelyn dedicates particularly extensive attention to the benefits of trees to their *communities;* the way the presence of
a grove, "arbour," or "close-walk" can offer not monetary value, but *social*
value, through their fruit and shade. Evelyn praises a supposed custom
from Frankfort that stipulated that "no young *Farmer* whatsoever is permitted to *marry a Wife,* till he bring proof that he hath planted, and is a
Father of such a stated number of *Walnut-trees;* . . . the Law is inviolably
observed to this day, for the extraordinary benefit which this *Tree* affords
the *Inhabitants.*" According to the "Ancient Law" of another area in Ger-
many, he goes on, "the *Borderers* were obliged to nurse up, and take care
of [the walnut trees]; and that chiefly, for their Ornament and Shade; so
as a man my ride for many Miles around that Countrey under a continued *Arbour,* or *Close-walk;* the *Traveller* both refreshed with the *Fruit* and
the *Shade.*"[33] In these examples, Evelyn ties the landlord's obligations to
his posterity to his obligations to the well-being of the "publick" by showing the ways that planting trees provides sustenance, succor, and security
over a long period of time.

Evelyn's extensive praise of such "Ancient Laws" as those he discusses
above suggests that he sees an inherent connection between a landlord's
ecological and social duties. Landlords, trees, and the present and future
well-being of family and society are persistently and intimately linked
throughout *Silva.* To cultivate and care for a community of trees, he suggests, is to cultivate and care for the socio-environmental community
writ large as well. But beyond that, the juxtaposition of language that
casts the trees as children of their planters and as the source of comfort and refreshment of inhabitants in *Silva* points to the ways that the
trees serve as key nodes in a network of interdependent, interspecies relationships. Calling the planters the "Fathers" of the trees, "obligated to
nurse [them] up and take care of them" underscores a communal—even
familial—relationship among humans and trees. In an extended simile
at the end of the second chapter of book 1, Evelyn compares cultivating
trees with raising children, emphasizing the need for time, patience, and
care with both: "from *Station, Sowing,* continual *Culture* and *Care,* pro-
ceed all we really enjoy in the World: Every thing must have *Birth* and
Beginning, and afterwards by Diligence and prudent Care, form'd

and brought into Shape and Perfection: Nor is it enough to cast *Seeds* into the Ground, and leave them there . . . great diligence is to be us'd in *Governing* them . . . after the same Method that our *Children* should be *Educated*; and taken care of from their *Birth* and *Cradle*."[34]

There is a clear georgic resonance in Evelyn's depiction of cultivation as equivalent to educating human children, and as in the *Georgics* it has the effect of making trees equivalent to children in terms of the "culture and care" owed to them. Unlike the *Georgics*, however, Evelyn's description focuses more on the issue of time, specifically the importance of the *continuity* of care that trees (and children) demand. The passage repeatedly insists that birth (or the casting of seeds) is merely the start of the long years of "Diligence and prudent Care" that are required to bring a tree (or child) to maturity. Thus, the long lifespan of the tree, and the fact that it represents the promise of the future, are also part of what makes Evelyn's comparison apropos, and part of what makes care for trees important. The fact that Evelyn likens the trees to the reader's own children is equally important, since it reinforces the parity of the landlord's duties of care for all of his posterity, human *and* nonhuman. Planting and caring for trees enacts the landowner's duty as a steward of posterity, broadly construed. The trees themselves, their long lives and the sort of care they demand, are central rather than incidental to Evelyn's case.

The main threat to England's forests and, by extension, to its natural and political stability is the same threat that shadowed the usufructuary ethos in chapter 1: shortsighted preoccupation with personal profit. Sometimes Evelyn portrays this shortsightedness as motivated by Cromwellian contempt for royal traditions, but more often he blames landlords' susceptibility to the get-cash-quick possibilities offered by the planting of grain. Evelyn imagines his landlord readers objecting that "'tis so long a day before these *Plantations*" of trees "can afford" any profit, as opposed to the "richest and most opulent *Wheat-Lands*," which pay out within, at most, a few years.[35] His main riposte to such objections is to point to the superior benefits trees offer to the public and posterity—to the idea, in other words, that the landowner has a responsibility to remember that others depend on his land and its use. (Land which, as Abraham Cowley writes in the introduction to his poem "The Garden" in the front matter of *Silva*, "we call [not very properly, but yet we call] our Own.")[36] Grain, Evelyn implies, is an attractive agricultural choice from the perspective of monetary profit, but ultimately a myopic and destructive one. It fails the test of proper use articulated by Swift in the last section: use that supports the landlord's "Brethren," present and future.[37] The core error of those who choose grain over trees, Evelyn implies, is that they see their land as a means to line their own coffers rather than as an inheritance as much of responsibility to others as of profit to themselves.

Evelyn was correct to be concerned about his readers' agricultural choices. Parliamentary efforts to restrain rapid agricultural "improvement" in late seventeenth- and eighteenth-century England mostly failed, a fact generally attributed to the opposition of common-rights holders to Parliamentary attempts to legislate new land- and tree-use limits, combined with the availability of trees that could be imported from England's colonies.[38] Given the widespread popularity of books like *Silva* that pushed an agenda of careful stewardship and long-term conservation, the failure of actual stewardship might seem surprising. On the contrary, I suggest that literary and devotional writing of the period strongly emphasizes the moral duties of the landlord to the earth and its dependents *because* of the failure of practical or legislative solutions. The already-extant discourse extolling the moral obligations of those with political and practical control of the environment intensified in late seventeenth- and early eighteenth-century popular devotional and literary writing because, absent practical or legal prevention of exploitation or despoilation, the usufructuary ethos offered a bulwark against the very real possibility of abuse and misuse of the nonhuman world by those who had the most control over it. It is precisely because of England's legislative failures that the task of guaranteeing a sustainable present and future became the moral and practical prerogative of the landlord, and why that prerogative had to be enforced through appeals to the duties of each being in a socio-environmental hierarchy, rather than through direct intervention by an earthly authority. Without a law that can be enforced, all that's left for writers like Evelyn is conscience.

Thus, when Evelyn decries the destruction England's woods have suffered at the very beginning of *Silva*, he is concerned with far more than naval timber reserves or the prerogative of the Crown:

> . . . it has not been the late increase of *Shipping* alone, the multiplication of *Glass-Works, Iron-Furnaces*, and the like, from whence this impolitick diminution of our *Timber* has proceeded; but from the disproportionate spreading of *Tillage*, caused through that prodigious havock made by such as lately professing themselves against *Root* and *Branch* . . . were tempted, not only to fell and cut down, but utterly to

extirpate, demolish, and raze, as it were, all those many goodly Woods, and Forests, which our more prudent Ancestors left standing, for the Ornament, and Service of their Country.[39]

It is tempting to read "*Root* and *Branch*" as a mere political metaphor, an artful arboreal pun on regicide and restoration. That pun is certainly a piece of the phrase's meaning. Yet as previous examples have shown, the resonance of "Root and Branch" as a moral touch point extends far beyond its royalist connotations. For one thing, the literal meaning cannot be dismissed: this is Evelyn's first articulation of his core premise, which is that planting trees is good, and cutting them down (without consideration and replanting) is bad, for practical environmental reasons. Those "professing themselves against Root and Branch" were thus literally against trees. That also means that they were against everything that trees represent and embody in *Silva*, which, as we have seen, includes public welfare and care and concern for the future. "Root and Branch" in that context takes on a new symbolic resonance: the literal roots and branches of trees that connect the metaphorical roots of the past with the metaphorical branches of the future through the tree's ongoing existence. To "extirpate, demolish, and raze" an actual forest is, in that framework, simultaneously to extirpate, demolish, and raze its embodiment of ancestors' prudence and its own future potential. Felling a forest destroys the living link between the past and the future by removing the trees that act as temporal media, and in felling those trees, the landowner violates his own role as the steward and generational medium of the environment. To do so in favor of "*Tillage*," moreover, wrongly reframes land use as a question of exclusively personal choice and profit on the part of the owner—a misconstrual of the land as "his" in an outright rather than a usufructuary sense.

Trees, Posterity, and Hierarchy in Finch's "Upon My Lord Winchilsea"

Anne Finch's "Upon My Lord Winchilsea's Converting the Mount in His Garden to a Terras" illustrates the usufructuary principles behind *Silva* in poetic form. The poem praises Finch's nephew, Charles, fourth Earl of Winchilsea, for his decision to restore the landscape of his estate to its previous state, removing a "mount" that had been built by his grandfather, Heneage, and replanting trees to re-create the grove that had previously existed on that site. Finch emphasizes the way her nephew reinstates the socio-environmental community that had been disrupted by Heneage and his mount, which Finch portrays as an aberration from the fulfillment of lordly duties to his human and nonhuman public and posterity.[40] The poem opens by positioning Charles as the bridge between his ancestors and his posterity, that signature temporal medial position of the usufructuary lord:

> If we those Gen'rous Sons deserv'dly Praise
> Who o're their Predecessours Marble raise,
> And by Inscriptions, on their Deeds, and Name,
> To late Posterity, convey their Fame,
> What with more Admiration shall we write,
> On Him, who takes their Errours from our sight?[41]

"Gen'rous Sons," according to Finch, preserve the connection between past and future by commemorating their ancestors. But she immediately complicates this by suggesting that an important piece of this intergenerational work might, paradoxically, involve strategic *erasures* of the mistakes and misdeeds of the past. The specific "Errour" to which Finch refers is the "Mount" of the title, which had taken the place of a "shelt'ring grove" of trees (23).

How can erasing the legacy of an ancestor protect the link between past and posterity? If, Finch answers, that erasure restores a socio-environmental legacy that had been wrongly destroyed. By razing the "Mount" and replanting the grove, Finch writes, Charles, the "wiser Offspring" of the family line, "does those gifts *renew*," returning the landscape and its denizens to their original, optimal state. The restored landscape is one where "trees" will "Florish" where "late the plough did passe," Finch goes on, and one that will "blesse the present, and succeeding Age" (75–76, 71). The restoration of continuity with the original environmental state of the land guarantees a stable, thriving present and future and ties intergenerational continuity and sustainability to trees specifically as embodiments and emblems. And, as in *Silva*, Finch implicitly contrasts the preservation of trees and posterity with the myopic preference for "the plough." Erasure of the newer, tree-free landscape acts as an ellipsis that reconnects two disconnected periods of familial and environmental history to one another, eliding the lapse in continuity. Forgetting acts as preservation of posterity. The new-old landscape is meant to make it possible to forget that any change had ever taken place.

While Finch's depiction of Charles affirms the usufructuary values of the good landlord, her depiction of Heneage illustrates the pernicious repercussions of a bad one for both posterity and the public. The fact that Heneage's destruction of the grove repudiates his environmental duties as landlord is evident not only in the fact that undoing his alterations is framed as a restoration of environmental and generational continuity but also in the fact that Finch calls pointed attention to the "untimely Fate," of the trees, which were "Sadly prescrib'd . . . a too early Date" (27–28). The problem is not just that they were cut down, but that they were cut down without consideration for the future repercussions of their absence. At the same time, the premature death of the trees "cause[s] a gen'ral Grief" and outcry, as "Some Plead, some Pray, some Councel, some Dispute" against Heneage's decision (29, 31). That "gen'ral Grief" expands the implications of his choice beyond his family and biological posterity to the broader community of people and organisms on which his choice has an impact. Finch makes a point of showing that resistance to the felling of the grove extended to Heneage's tenants: "The very Clowns . . ./ Refuse to strike" down the trees, "nor will their Lord obey" until he himself strikes the first blows (37–38). The "gen'ral," instinctive resistance to destroying the trees among all those who depend on the estate signals the deep significance of the environmental stability of the estate as a function of the community's social cohesion. The "clowns'" refusal to obey "their Lord" is a threat to the proper social order of the estate. But the blame for it lies with Heneage, rather than with his tenants, as their revolt has its roots in their deep awareness that what they are being asked to do is in itself wrong.

In fact, the clearest sign that something is awry with Heneage's lordship comes several lines earlier, when Finch sighs that his tenants objected to his orders "Alas in vain, where Pow'r is Absolute" (32). The power of the human landlord in the usufructuary ethos, as we have seen, was never properly absolute. Like his ownership, the lord's power is insistently framed throughout the period as temporary, limited, and revocable. Riches, and the power to use and benefit from them, exist to be stewarded for the benefit of all, not exploited for the enjoyment of one. That Heneage considers his prerogative "Absolute" is therefore a signal that he has violated the "trust" he has been given as the lord of the estate, a violation that his tenants seem to sense, but are unable, ultimately, to resist. In fact, Heneage's own actions lead them astray: after he lands a "leading stroke, / [He] by Example does their Rage provoke" against the innocent grove (39–40). His shortsighted, selfish insistence on chopping down the trees leads others to embrace his anarchic, anticonservationist mindset, only for proper social relations to be restored along with the trees later on. Landlordly duty and social stability are thus tightly intertwined with the conservation and importance of trees.

Understanding this fact helps to resolve the apparent tension between the political and environmental levels of Finch's poem, and to elucidate an aspect of the cultural significance of the stewardship of trees that a too-tight focus on trees' allegorical political significance has previously occluded. The tendency to jump straight from trees to oaks to Stuart symbolism in scholarship on Finch has tended to distort this poem by presuming that the royalist symbolic program is always the most prominent level of meaning, and that since Finch is criticizing Heneage, he must in some way represent her political enemies. Finch's Stuart loyalties are indisputably a fundamental aspect of her poetry; I do not wish to dismiss readings that draw out their importance. But focusing exclusively on trees' political significance conceals the other ways trees signify, particularly through biological traits such as longevity. In the case of "Upon My Lord Winchilsea," the presumption that the trees she describes must be oaks and therefore Stuart symbols has led critics to presume that Finch's critique is subtextually aimed at some political adversary, rather than at the problem of failures of stewardship among landlords more generally. Wes Hamrick, for example, persuasively ties "Upon My Lord Winchilsea" to a literary tradition reaching all the way back to Ovid in which bad landlordship is represented by the act of cutting down trees. But his conviction that all trees in Finch automatically are oaks, and therefore symbols of her royalist loyalties, leads him to conclude that "the literary tyrant who cuts down trees or disturbs the forest is never far removed" in Finch from "Cromwell, the Parliamentarians, or William III."[42] The problem is that that claim runs counter to Hamrick's own evidence that the connection between conservation of trees and good lordship long predates any of those political figures or the political conflicts of the late seventeenth century.

References to bad landlords who chopped down trees were not automatically veiled partisan references per se, but rather references to a broad and complex fabric of moral thought that weaves together political, social, and environmental significance. Nicolle Jordan, in her groundbreaking reading of the poem's environmental subtext, points out that Finch's poem "provides an intimate encounter" with the ways "morality entered into the discourses of forestry and improvement" in the late seventeenth century, further noting that the poem "leaves readers with an acute awareness that paternalism occurs in both benevolent and malevolent forms" and that those forms have significant environmental repercussions.[43] Yet her presupposition that the landlord figures in the poem are emblems of Stuart monarchy and thus, because of Finch's royalism and affinity for patriarchal hierarchism, above reproach leads Jordan to struggle to account for the ways the poem appears both to support and to critique Tory ideology. She ultimately concludes that the poem "helps us to understand the circumstances that fortified—and politicized—the imbrication of morality, land management, and political legitimacy . . . despite the absence of a coherent politics."[44] What my reading of Finch and Evelyn suggests is that the explicitly partisan frames that have been used to interpret the poem's environmental ethos are political red herrings, creating the appearance of political incoherence by locating the confluence of politics and environmentalism in the wrong place. My reading sets aside the well-explored program of political symbolism associated with the oak to resurrect some of those other ways that trees signified for early modern writers, in particular how they functioned as a key literary trope of the usufructuary ethos through their embodiment of mediality and accountability to the public and posterity. In doing so, it becomes possible to see that by praising Charles's stewardship of public and posterity and condemning Heneage's selfishness and shortsightedness, Finch is participating in and perpetuating a form of idealized conservationist landlordship that was not exclusively either Tory or Whig, but that *was* exclusive to the landowning class.

Nicolle Jordan helpfully reminds us that the "proto-environmentalist sentiments" found in Evelyn and Finch "precede and deviate from the modern tendency to associate environmentalism with struggles for social reform," and thus provide a "salubrious reminder of the ongoing transformation and instability of political labels."[45] I suggest that we can go a step beyond this: this arguably most prominent strain of long eighteenth-century conservationist thought is fundamentally grounded in a theory of social, political, and natural hierarchism that put landowners in charge of protecting the environment, but—for all its obvious Tory resonances—it was not in itself fundamentally partisan at the time. Rather, it was a widely held belief that found expression in various manners among authors of apparently disparate political beliefs. As Jordan herself points out, Finch's poem "endorses an ethos of stewardship to which both Whigs and Tories are laying claim at the end of the seventeenth century"—a signal that perhaps the trees in the poem are never marked specifically as oaks because they are not intended as part of the symbolic program of the Jacobite cause. This is a poem not about trees as political symbols, but about trees as ecological linchpins in a broad and interconnected system of socio-environmental bonds, whose fundamental underpinning is, as Jordan terms it, a "paternalistic natural order."[46] But Finch's shoring up of that order in "Upon My Lord Winchilsea" is not necessarily a sign of her Toryism so much as of her implicit acceptance of the moral, social, and environmental terms of the usufructuary ethos.

One final point must be made about the "paternalism" of the usufructuary environmental ethos that Evelyn and Finch endorsed: not only did it elevate the moral and environmental duties of the landlord but, correspondingly, it demoted the environmental role of the lower class. Evelyn repeatedly insists that commoners cannot be trusted with long-term stewardship. On the third page of "To the Reader," Evelyn advises that landowners "not easily commit themselves to the *Dictates* of their ignorant *Hinds* and *Servants*, who are (generally speaking) more fit to Learn than to Instruct." Landowners, he goes on, ought to look to the examples of the lords who have come before them, or even back to classical and biblical literature, rather than to the lower classes. Later, he avers that commoners cannot be trusted to act beyond their own shortsighted self-interest. In order to achieve the ecological and economic aims *Silva* lays out, Evelyn insists, "his Majesty must assert his *Power*, with a firm and high Resolution to reduce these Men to their due *Obedience* and to a necessity of submitting to their own and the public utility . . . ; while some person of trust and integrity . . . regulate and supervise" the management of new woodlands.[47] Likewise, Heneage's first blow against the grove was especially heinous because it set a new precedent for his laborers, who, in Finch's poem, follow the moral and environmental example of their landlord. Decisions about use and long-term planning are the purview of the landowner, according to Evelyn, not only because they are legally the only ones allowed to make such decisions but because by the very nature of their roles and identities they are the only ones capable of it. Foresight, broad concern for the well-being of human and nonhuman dependents, and a sense of duty and accountability to future generations and to God as the giver and ultimate owner of their worldly gifts—these traits define the landlord specifically and exclusively, and in turn his tenants are defined by their lack of those qualities. That lack is a piece of the hierarchy that undergirds the usufructuary ethos: every rank in the socio-environmental chain has its role, and it must stick to that role.

The equation of conservationist with landlord creeps into claims for Evelyn's proto-environmental bona fides even in contemporary scholarship where the author focuses on the environmental side of the question over Evelyn's particular social and political context, and on periods much later than the late seventeenth century. James C. McKusick, for example, argues that Evelyn "foreshadows the development of a conservationist ethic in the management of forests and wildlands throughout the English-speaking world," then goes on to tie an 1827 Sir Walter Scott article on forestry to Evelyn via Robert Southey's praise of Evelyn in an 1818 *Quarterly Review* essay on Evelyn's *Memoirs*.[48] The connection is persuasive, but the terms of Southey's praise for Evelyn's impact on English landscapes is telling: Evelyn saved English forests, Southey writes, "by awakening the land-holders to a sense of their own and their country's interests."[49] The long-term sustainability of the forests and the populations who rely on them is still the desideratum for Southey—and the landowner is still the figure through whom, and only through whom, that desideratum can be achieved. As in Finch's case, Evelyn's defense of this hierarchy is not motivated by his identification with royalist or Tory ideology per se, but rather by a more fundamental and less partisan—though no less political—belief in the ontologically hierarchical tenets of the usufructuary ethos.

Philips's *Cyder* and the Usufructuary Lord

Like *Silva*, John Philips's 1708 georgic poem *Cyder* digs into the cultivation and care of trees in order to articulate a socio-environmental vision that places the landlord—and more specifically, his uses of his land and wealth—at the center of the environmental and political stability of England. Juan Christian Pellicer has detailed Philips's reliance on the section of *Silva* dedicated to apple trees, "Pomona," but an examination of the socio-environmental role Philips assigns the landlord in his poem reveals a debt to Evelyn's *Silva* that goes well beyond silviculture.[50] This is an important point to pause on not just to connect Philips to Evelyn through a similar politico-environmental vision, but because, as the first English poetic *adaptation*—as opposed to *translation*—of the georgic, *Cyder* brings the concepts and concerns of the usufructuary ethos to the classical genre. *Cyder* demonstrates how the English georgic became one of the key genres of poetry through which writers strove to accommodate and articulate the relationships among ethics, politics, and environment amid shifting economic, environmental, political, and social conditions. The fact that Philips adapts the Virgilian georgic to defend the place of the landlord in early eighteenth-century Tory political economy reveals not merely Philips's of-the-moment partisan and ideological commitments but also the way the formal georgic came to be the genre through which the key social and environmental values of the usufructuary ethos were adapted to fit the rapidly changing political, economic, and ecological worlds of the eighteenth century.

This section explores the ways that values of the usufructuary ethos and the socio-environmental landlord shaped the English georgic. Virgil's *Georgics* offered classical authority—both moral and poetic—to the exploration of the intersections of the human and nonhuman worlds through labor and cultivation.[51] Furthermore, the way Virgil's poem interweaves politics and environment, and reinforces the moral importance of human beings' medial position, was congenial to the goals of long eighteenth-century poets. Philips's poem represented an early attempt to recruit Virgil's classical authority in support of early eighteenth-century English land ownership and use by poetically linking the two. Philips also recruits the georgic's moral authority in order to critique contemporary agricultural and economic developments he was wary of. Whereas Virgil's original poem contrasted the virtuous farmer and the rapacious and/or luxurious urbanite, *Cyder* turns its focus to the more subtle question of what constitutes morally defensible motives for, approaches to, and uses of the "gains" made possible by cash-crop agriculture. Philips's georgic thus merges classical topoi with the usufructuary ethos in order to defend and promote that particular socio-environmental order against competing social, environmental, and economic values. His project is by its nature a simultaneously environmental and political one, concerning both the proper use of and care for the land, and a defense of the ways land and power were distributed in England.

Yet as was the case with Finch, critical responses to *Cyder* have thus far tended to divorce its political and environmental dimensions. Most recent work has approached it as a poem "steeped in politics, in a ... partisan and specific sense" as well as in the sense of functioning as an ideological tool of the various figures involved in its production and distribution.[52] As a result of that focus, however, these readings tend to treat *Cyder's* depictions of agriculture and nonhuman nature either as generic tropes or as emblems of political or partisan positions. Philips and his contemporaries, according to Pat Rogers, were "working in a vein of politicized georgic, where the traditional materials of rural poetry are reprocessed in the light of topical concerns, party loyalties, and competing visions of English history. . . . [Georgic] at this time took an abrupt turn in the direction of public themes and patriotic subtexts."[53] The underlying assumption here is that such a turn toward "public themes and patriotic subtexts" made the nonhuman beings depicted in eighteenth-century georgics mere emblems of the Tories' political positioning of themselves as the opponents of Whiggish promotion of the new money economy. My reading of *Cyder* will remedy this political/environmental split by showing the ways Philips's political and environmental landscapes mutually reinforced one another.

In *Cyder's* adaptation of Virgil's Old Man of Tarentum episode from book 4 of the *Georgics*, Philips interpolates the specific socio-environmental frameworks of the early eighteenth-century usufructuary ethos into his adaptation of the georgics. He contrasts his idealized usufructuary landlord with two portraits of the behavior of bad landlords, each of whom exemplify potential abuses of land that arise from for-profit agriculture. Philips's version of the prudent, diligent, responsible Old Man of Tarentum is

> A frugal man . . .
> Rich in one barren Acre, which, subdu'd
> By endless Culture, with sufficient Must
> His Casks replenisht yearly: He no more
> Desir'd, nor wanted, diligent to learn
> The various Seasons, and by Skill repell
> Invading Pests, successful in his Cares. . . .[54]

Like the classical georgic farmer, Philips's "frugal man" is "diligent to learn" the needs, habits, and demands of the land he dwells on and cultivates, and is engaged in the productive conflict with nature to survive and thrive that is distinctive of georgic labor. Philips's "frugal man" concerns himself not with surplus or profit, but rather with maintaining a ready supply of resources for present and future use. Given the emphasis throughout

Cyder on the title drink's ties to hospitality and communal celebration (a few pages later, Philips lauds "he with bounteous Hand / [who] Imparts his smoaking Vintage, sweet Reward / Of his own Industry"),[55] the "frugal man" embodies the focus on public and posterity characteristic of the usufructuary landlord, merging Virgil's classical ideals with the similar, though not identical, eighteenth-century ideal.

Philips juxtaposes the frugal man's habits with two examples of what a landlord/cider producer ought *not* to do, both of which represent ways to maximize personal profit by denying others their due support. This move introduces a crucial motif into the eighteenth-century English georgic: the ethics of the use of riches and the anxieties and challenges generated by the growing association between agriculture and monetary profit. First, he warns, "No heterogeneous Mixtures use, as some / With watry Turneps have debas'd their Wines, / Too frugal,"[56] a straightforward condemnation of stretching profits from the sale of apple cider by mixing it with cheaper juice from turnips. The second warning touches on an issue of political controversy in the early eighteenth century, the justification of tithing to support Anglican clergy. Philips, as one would expect of a staunch Tory, is in support of tithes, but what's most significant about this passage is the way he connects the issue of tithes to what preceded it through the problem of avarice:

> Nor let thy Avarice tempt thee to withdraw
> The Priest's appointed Share; with cheerful Heart
> The tenth of thy Increase bestow, and own
> Heav'n's bounteous Goodness, that will sure repay
> Thy grateful Duty: This neglected, fear
> Signal Avengeance, such as over-took
> A Miser, that unjustly once with-held
> The Clergy's Due; . . .
> Be Just, be Wise, and tremble to transgress.[57]

One common argument in support of tithes around the time Philips was writing *Cyder* drew on the belief that human beings were the usufructuaries of God, and that therefore it belonged to God to determine what portions of his property rightly belonged to whom. The dean of Norwich Humphrey Prideaux (1648–1724) summarized the argument in 1710: "Ministers' Maintenance is God's property, and . . . we rob him, whenever we detract from them any part of that Maintenance. . . . [God] gave

[clergy] these Tithes to receive and enjoy them in a usufructuary tenure under him, as their Wages from him."[58] Philips draws on these same
theological premises for his warning that failure to "repay / Thy grateful
Duty" of paying tithes is tantamount to misappropriation of God's property, and a failure of the duty that requires landlords to fulfill their duties as the medial, usufructuary lords tasked with distributing God's gifts
as they were meant to be distributed. The fact that "Avarice" inspires this
transgression—whether avarice to keep the monetary profits from selling
the extra produce, or avarice to keep that produce for one's self—points to
greed as the number one threat to the values and socio-environmental vision of the usufructuary ethos, which, for Philips, is in turn bound up
with defending the signature Tory issue of tithing. There is no precedent
in Virgil's *Georgics* for either turnip juice or tithing, but by embedding
these concerns in the center of *Cyder*, and tying the problems of avarice
and tithes through negative comparison to his version of the recogniz-
ably georgic Old Man of Tarentum, Philips adapts the formal georgic to
contemporary English questions of the ethics of use.

Philips's use of Virgil's form and topoi reshaped the English georgic
into the poetic genre in which the tensions among use, politics, morality, and the shifting material and conceptual conditions of agricultural
production could be staged. It also enabled Philips to bolster his praise
of particular aristocratic figures by poetically linking them to the moral
and civic authority of the *Georgics*. The partisan political significance of
the people Philips chose to praise in his encomiums on landlords in book
1 has been well established by scholars like Pellicer and Rogers, but the
specific terms in which he couches that praise are worth revisiting in light
of the usufructuary ethos. Like other writers we have encountered so far,
Philips emphasizes the landlord's role as the guarantor of the well-being
of public and posterity. He praises his chief patron for the poem, Robert
Harley, then secretary of state, in familiar terms: as "Sollicitious of public
Good"—in this context probably a reference to his political service—and
also, more importantly, as one who "gathers but to give, / Preventing
Suit."[59] The latter bit of praise gestures back to the usufructuary ethos's
conception of the lord as the conduit through which heaven's "Blessings"
and "Nature's Gifts" (to use to Philips's own language) pass from God
to his creatures, and from past to present to future. That Harley sup-
posedly gives without having to be prompted implies that he is aware of
and honors the duties incumbent on his position, and the medial socio-
environmental role he occupies.

Other figures receive praise that similarly situates them in terms of their medial roles with regard to posterity or the public. Philips contextualizes his praise of James Brydges, eighth Lord Chandos, in terms of his cross-generational links to his ancestor, Sir John Brydges, who "transmits Paternal Worth" to his descendant.[60] Sir Thomas Thynne, first Viscount Weymouth, keeps his

> hospitable Gate,
> Unbarr'd to All, invites a numerous Train
> Of daily Guests; whose Board, with Plenty crown'd,
> Revives the Feast-rites of old. Mean while His Care
> Forgets not the afflicted, but content
> In Acts of Secret Goodness, shuns . . . Praise. . . .[61]

Weymouth embodies one who "gathers but to give," exemplifying the praise that Philips goes on to give Harley. Both Chandos and Weymouth, furthermore, are framed as figures through whom possession passes; as stewards, and transmitters, rather than outright owners. The passage from which these encomiums come—a series of flattering portraits of Tory and nonjuring men—is perhaps the most politically charged of the whole poem. Yet even here, Philips relies upon appeals that frame their moral and social eminence as the product of their adherence to usufructuary ethics of power and possession.

It is undeniable that praise for passing down "Paternal Worth" and of carrying on "Feast-rites of old" invoked nostalgic values of lordship in service of politicized Tory attempts to romanticize landed wealth and power. But the fact that even *Cyder's* most directly partisan passages exist at the center of a *georgic* poem binds their political significance to questions of proper use of and responsibility to the land and its inhabitants. In both the georgic and the usufructuary ethos, the political and environmental are inseparable. A vision of the political good is a vision of the environmental good. By imbricating the usufructuary socio-environmental ethos with the formal georgic's didacticism and preexisting cultural authority, Philips solidified the formal English georgic as the genre through which eighteenth-century writers processed and attempted to reconcile the material, economic, and political developments of their day to traditional socio-environmental ethics of use that still, for a while at least, held cultural authority.

Virgil, Dryden, Philips, and the Slide from Subsistence to Profit

The most significant new focus Philips introduced to the georgic was *Cyder's* sustained concern with the problem of the use—and misuse—of riches, meaning here both "Nature's Gifts" and the monetary wealth that agriculture could generate. As in Evelyn's *Silva*, avarice and wealth shadow Philips's depictions of agriculture. That concern reflects the fundamentally different relationship among humans, environment, and economics underlying the English georgic compared to Virgil's original poem. Unlike the world of Virgil's *Georgics*, eighteenth-century England was a time and place when agriculture could and did produce wealth and luxury for landlords, and for all their nostalgia, that was a fact that Philips's Tory patrons benefited from as much as any Whig did. In order to understand the full significance of that new focus, and to begin to illustrate the disparity between the Virgilian georgic ethics English poets attempted to recruit and the context they recruited it for, we must return briefly to Virgil's original *Georgics*. The imperative to provide for the future survival of one's family and dependents, I argue, provides the ethical foundation for the often violent, controlling, or combative behavior humans and other animals engage in in Virgil's *Georgics*. The emphasis on survival is crucial to Virgil's political and environmental vision, and provides the basis for the contrast between the violence of the virtuous farmer and the violence of the greedy, luxurious urbanite. With Dryden's 1697 translation, however, that stark Virgilian ethical contrast began to shift. In both Dryden and Philips, though the anxiety about profit is present, so too is the notion that *some* forms of profit can be good. The problem became how to adjudicate "good" forms of profit from "bad."

Georgics are famously and generically preoccupied with the best ways to reap a plentiful, reliable harvest from a recalcitrant natural world. In Virgil's *Georgics*, that goal is always couched in terms of subsistence. In book 1, Virgil depicts the labor of agriculture as a literally life-or-death battle: "Therefore, unless your hoe is ever ready to assail the weeds, your voice ready to terrify the birds, your knife to check the shade over the darkened land, and your prayers to invoke the rain, in vain, poor man, you will gaze on your neighbour's large store of grain, and you will be shaking oaks in the woods to assuage your hunger."[62] All that assailing, terrifying, and checking is necessary, Virgil says, because if you do not do it, your crop will fail, or be eaten, and you and all your dependents will starve. Virgil's poem stages the conflict among living creatures as a zero-sum game: either you get the food or I do, and whoever loses will suffer. That constant tension among living beings over the means of subsistence manifests as a dialectical movement in human beings' relationship to nonhumans between conflict and care.[63] Book 2, for instance, depicts cattle first as dangerous pests, then as valued family members. Virgil urges farmers to "weave hedges, and keep out all cattle, chiefly while the leafage is tender and knows naught of trials, for besides unfeeling winters and the sun's tyranny, ever do wild buffaloes and pestering roes make sport of it; sheep and greedy heifers feed upon it. No cold, stiff with hoar frost, no summer heat . . . has done it such harm as the flocks and the venom of their sharp tooth, and the scar impressed on the deep-gnawed stem."[64] Virgil describes both wild and domesticated cattle as menaces to young crops, their poisonous teeth capable of inflicting permanent damage. They are mindlessly appetitive, their greed driving them to despoil the fields if they are not completely kept out. What's more, they receive both human and divine punishment for their crimes: "for no other crime," Virgil writes, is "a goat slain to Bacchus at every altar."[65] Cows and goats are in direct and violent conflict with humans at this stage of the growing season.

By the end of book 2, however, the relationship between humans and their cattle shifts from conflict back to care, and the fight to protect the young shoots of grain against cattle turns out to have been motivated by the foresight that that grain will be needed to feed both cattle and humans in the winter. The farmer's springtime parsimoniousness enables him later "gladly" to bring "them their food and provender of twigs, . . . closing not [his] hay lofts throughout the winter." The battles the farmer fought against animals earlier in the year ultimately enable him to provide "sustenance for his country, and his little grandsons, [and] for his herds of cows with faithful bullocks": for every living being, in other words, in his extended socio-environmental community.[66] That sense of reciprocity and care between farmer and cattle is even more pronounced in Dryden's 1697 translation, in which he renders the previous passage thusly: "His Wife, and tender Children to sustain, / And *gratefully* to feed his dumb *deserving* Train."[67] The idea that the farmer is *grateful* to his cattle, who earn their share by virtue of their cooperation with humans in the process of producing the harvest, and that the cattle receive their food as just compensation for their help in producing that food, is more muted in Virgil. But in both texts, the farmer's concern for the well-being of all the living creatures for whom he is responsible motivates both his care for the cattle during the winter and his conflict with them during the spring. To be a living, eating, thinking creature in the *Georgics* is to be required constantly to remember that the earth and its produce are fair game for all to use, that you are a key link in the chain that preserves present resources for future use, and that you are accountable for ensuring that all does not "degen'rate still to worse," as the entropy of the post–Golden Age world is always in danger of doing. The stakes in Virgil's *Georgics*, therefore, always boil down to survival—not profit.

Virgil's critique of dissipated and corrupt urbanites in book 2 highlights the extent to which the poem severs wealth, power, and profit from agriculture and contrasts the havoc sown by the pursuit of the former with the peace and plenty supplied by the latter. His head turned by greed for luxury and wealth and power, one member of the urban military elite "wreaks ruin on a city and its wretched homes, and all to drink from a jewelled cup and sleep on Tyrian purple; another hoards wealth and gloats over buried gold. . . . They steep themselves in their brothers' blood and glory in it; they barter their sweet homes and hearths for exile and seek a country that lies beneath an alien sun."[68] The urban elites' separation from "home and hearth" corresponds directly to decay in social and civic values, as they "barter" them for the promise of a "jewelled cup" or "buried gold." Virgil even goes so far as to imply that such separation leads not merely to neglect of the home and the familial and social structures that underpin life, but to civil war, steeping in "brothers' blood."

Here we also see an example of the political merging with the environmental as well, for such greed for wealth and power leaves its marks in the land itself, disrupting agriculture for far longer than merely the length of one war. In Kevis Goodman's eloquent phrase, the *Georgics* frames the "present as the future's past."[69] The poem constantly tries to assess possible courses of action from the perspective of their potential outcomes, seeing the choices of the present moment as ways to secure—or fail to secure—survival in an uncertain world. The most famous example of the notion of the present as the future's past comes at the end of book 1, when Virgil imagines the refuse of the civil wars being uncovered by future generations of farmers: "Yes, and a time will come when in those lands the farmer, as he cleaves the soil with his curved plough, will find javelins corroded with rusty mould, or with his heavy hoe will strike empty helmets, and marvel at gigantic bones in the upturned graves."[70] The present war rears up from the earth as a specter of the future's violent and unproductive past. Weapons and bones disrupt the farmer's attempts to get on with the work of farming, hinting at the way the past continues to shape the future, for better or worse. The wars may be over, but they still interfere with the work of survival.

The lines that follow drive home the long-term effects of neglecting agriculture in favor of war: "so many wars overrun the world, sin walks in so many shapes; respect for the plough is gone; our lands, robbed of the tillers, lie waste, and curved pruning hooks are forged into straight blades."[71] These lines clearly lament the effects of war while the war is going on. But the previous lines' relics of war turning up again in the future suggest that impoverishing the world now will have repercussions for future generations, who will reap the consequences. The fact that the fields "lie waste" and plains are empty during war means that the struggle to recover and thrive in the postwar world will be that much more difficult, survival that much more uncertain. And since human and nonhuman well-being is so utterly intertwined in the *Georgics*, the consequences of the human decision to go to war, a decision that is fundamentally political, affect the welfare of nonhumans as well.

Yet as the urban and military elite strive after wealth and glory, Virgil's farmer is still fighting to produce food to support everyone—including the elite. His proper care for and attention to the natural world in spite of interruptions by politics and greed is the cornerstone of not only his human and nonhuman dependents' survival but his country's as well. The issue of the farmer's monetary profits never arises. Wealth and power, in Virgil, are gotten through war or conquest, not through agriculture. The threat that wealth poses to society and to the environment in the *Georgics* is more one of the getting than of the use. There are those who "hoard wealth and gloat over buried gold," but the direct threat their avarice poses is attenuated in comparison to those who "wreak ruin" in pursuit of riches. The miser and the marauder are equally disconnected from agriculture, which is the source of physical and social survival; it is not a source of wealth or supremacy. The moral locus of agricultural labor in Virgil's *Georgics* lies in the way that agricultural labor underpins the interdependent webs of subsistence and social survival among all creatures.

The first signs of change in the moral relationship between earth and economy in English georgic appear in Dryden's 1697 translation. For all that Dryden preserves features of Virgil's environmental vision such as the dialectical movement between conflict and care, one significant way his translation departs from the original is in its depiction of trade and, by extension, the relationship between agriculture and wealth. A brief look at one key moment from Dryden's *Georgics IV*, the allegory of the beehive, will demonstrate the translation's connections to the usufructuary ethos, illustrate the unique way that Dryden's translation merged the social and the environmental, and begin to illustrate the subtle shifts that occurred between Virgil's *Georgics* and eighteenth-century English georgics in their depictions of the relationships among environment, labor, and wealth. Describing summer activity in the hive, Dryden writes:

> Exalted hence, and drunk with secret Joy,
> Their young Succession all their Care employ:
> They breed, they brood, instruct and educate,
> And make Provision for the future State:
> They work their waxen Lodgings in their Hives,
> And labour Honey to sustain their Lives.[72]

The contemporary political resonances here are clear and have been well canvassed. Listening with an ear attuned to the usufructuary ethos, however, elucidates the way Dryden's rendering of this passage merges sociopolitical and environmental concerns. Dryden parallels "making Provision for the State" by raising and educating their children—figuratively storing up a "provision" of future citizens and leaders to preserve a political "state"—with literally storing up provisions of "Honey to sustain their Lives." The two kinds of labor—raising children, storing honey—are linked by the way they both exemplify the usufructuary ethos's characteristic temporal mediality: the labor of the present is the bridge that connects a politically and ecologically stable past with a politically and ecologically stable future. That mediality was interpolated into the passage by Dryden; there is no such explicit concern for posterity in this particular passage of Virgil (a sign, perhaps, of how important temporal mediality was to early modern culture).[73]

A little later on, Dryden plays up the commitment to public and posterity that Virgil attributes to the bees, in language that invokes the political and economic concerns of 1690s England: bees are

> common Sons, beneath one Law they live,
> And with one common Stock their Traffick drive. . . .
> Mindful of the coming Cold, they share the Pain:
> And hoard, for Winter's use, the Summer's gain.[74]

Bees even have division of agricultural labor from civic labor, according to Dryden: "The youthful Swain, the grave experienc'd Bee: / That in the Field; this in Affairs of State / . . . / Their Toyl is common," and both contribute equally, though differently, to the present and future survival of the hive,[75] as the different ways of "making Provision" did in the earlier example. The political and the agricultural, the fabric of the social and the ecological community, therefore, are mutually constitutive. Temporal mediality and the interconnection of politics and environment are consistent themes throughout Dryden's *Georgics*, but at perhaps their most intensified in book 4, where the bees become emblematic of an ideal socio-environmental polity.

Yet while Dryden follows Virgil's original text in emphasizing subsistence as the key driver of bees' activity, his chosen language occasionally hints at his very different economic context. The very first periphrasis Dryden interpolates for bees is "trading Citizens,"[76] which, combined with phrases like "with one common Stock their Traffic drive," points to a subtle but crucial distinction between Virgil's farmers and the landlords of Dryden's age, a time when exporting grain had become a major source of income—even wealth—for English landowners. Recall that Evelyn's projected opponents in *Silva* were not just the wasteful republican landholders who had cleared woodlands during the Interregnum, but also contemporary landlords who saw a quicker profit to be gained by clearing woodland to grow grain than in preserving those woods for the future.[77] The so-called Economic Revolution was in full swing by 1697, opening up new avenues for wealth both in the expanding colonial world and in England, as projects to "improve" estates through enclosure and the draining of fens and reclaiming other "waste" lands opened up new sources of land to be farmed for profit. Dryden's "trading" bees, with their "stock" and "Traffic," introduce into the georgic an explicit connection between agriculture and wealth that was largely absent from Virgil. The change reflects the various cultural, economic, political, and material circumstances that shaped English relationships to land and farming. The fact that he wove those changes into his translation reveals Dryden's desire to reconcile classical georgic values with modern mercantile ones, to claim continuity with the past.

Like Dryden, Philips transformed the georgic topos of the bee to address the specific concerns of an agricultural world inextricably entwined with the commercial world, but—as befits the fact that his was an adaptation of the mode rather than a translation—he took it a step further.

Departing from Virgil's subsistence-farming bees, and building on Dryden's industrious, trading hives, *Cyder* introduces a new insect: the rapacious wasp. Philips unfolds an extended metaphor comparing wasps to avaricious men, setting up a theme of argument and admonition about the socio-environmental implications of good versus bad use and profit that will recur in poetry through rest of the eighteenth century. In *Cyder*, wasps are a pest species that threaten the apple crop. They "drain a spurious Honey from" the apple trees for "Their Winter Food":[78] the wasps getting their winter food, Philips implies, means that the humans will be denied their winter supply of apples and cider. Philips suggests a method for eradicating wasps by luring them in with the scent of thick, sticky "*Moyle,* or *Mum,* or *Treacle's* viscous Juice," which will trap their feet and bind them "'till Death / Bereave them of their worthless Souls: Such doom / Waits Luxury, and lawless Love of Gain!"[79]

Cyder's wasps are, in one sense, a classic example of the georgics' zero-sum struggle between human farmer and nonhuman creature. But Philips's metaphor attributes the wasps' behavior not to subsistence, nor to a legitimate struggle for survival, but rather to "Luxury, and lawless Love of Gain." In their pursuit of "luxury" and "gain," the wasps steal and destroy the fruit that supports the natural community as a whole. In Philips's depiction, wasps are separate from the interdependent web of lives that keep the orchard and its various denizens alive. Wasps see its produce only as something to be appropriated. Their greed and gluttony, Philips implies, is an existential threat to the very fabric of the community. While Philips's condemnation of the wasps' gluttonous "luxury" would arguably fit with Virgil's condemnation of those who destroy cities to "drink from a jewelled cup," the reference to a "lawless Love of *Gain*" indicates more modern anxieties about the moral and material repercussions of seeing the world as a source of personal profit rather than as an interconnected—if hierarchical—community of beings collaborating for the benefit of the public and posterity.

The rhetoric of the publicly, intergenerationally, environmentally responsible landlord persisted in devotional and literary writing in England well into the eighteenth century, in similar (though shifting) forms. Writing praising good uses of riches and satirizing bad remained common from the Restoration throughout the Augustan period, reflecting both the persistence of the usufructuary environmental ethos and a growing anxiety that it was not being adhered to. The problem—of which writers like Evelyn, Finch, and Philips were well aware—was that those who had the power to make large-scale land-use decisions (land-lords) were under no direct legal or earthly obligation to make *good* choices. Like Heneage Finch, any landlord could choose to flout his obligation to public and posterity at any time. Philips's wasp, finally, reveals an underlying awareness of and anxiety about this exact problem: how to distinguish between good and bad uses of "riches" in the face of growing economic temptation. After all, the fact that there is a "law*less* Love of Gain" suggests that there must also be a law*ful* love of gain. Philips condemns not improvement or monetary profit from agriculture per se, but rather the avariciousness he attributes to those who pursue personal profit at the expense of their local community and their nation. If one can do it *right*, if one can articulate how one is still fulfilling the socio-environmental demands of the usufructuary ethos, there is room, morally, for profit, even for pursuing profit. It is a delicate balancing act that was picked up by poets later in the century who sought to uphold the usufructuary ethic of use that still held authority—and in those poems, beginning with Pope's, we shall see the tensions that remain somewhat muted in Evelyn's, Finch's, and Philip's work come to a head.

3

Pope and the Usufructuary Ethics of the "Use of Riches"

Almost exactly halfway through the *Epistle to Burlington*, the first of Pope's two moral epistles on the "use of riches," there is an anecdote about fathers, trees, and intergenerational connection that resonates strikingly with Pope's seventeenth-century predecessors. First, Pope describes the ideal father-landlord, Sabinus, and his relationship with the trees on his estate:

> Thro' his young Woods how pleas'd Sabinus stray'd,
> Or sat delighted in the thick'ning shade,
> With annual joy the red'ning shoot to greet,
> Or see the stretching branches long to meet![1]

The repetition of present participles—"thick'ning," "red'ning," "stretching" —to describe the source of Sabinus's pleasure emphasizes that his delight in the woods stems from the fact that they physically enact connections among past, present, and future. Sabinus's ancestors sat under these trees, and they will support the lives of his children and grandchildren. Likewise, the fact that his is an *"annual* joy"—that is, one that he repeats year after year—indicates that memory is as much a part of Sabinus's experiences as projection. He honors the past and plans for the future, and, like Ulysses's father in Evelyn's *Silva* and the good lords in Finch's "Upon my Lord Winchilsea," the trees not only symbolize but embody that temporal connection. They are his living connection to posterity, and it is the particularities of trees' different lifespans and timescales from humans' that make that possible. Recognition of

and gratitude for that are part of what makes Sabinus a good landlord, along with the care and protection of his woods that his gratitude
inspires.

But all of Sabinus's care and planning is destroyed by his son's "fine
Taste":

> His Son's fine Taste an op'ner Vista loves,
> Foe to the Dryads of his Father's groves,
> One boundless Green, or flourish'd Carpet views
> With all the mournful family of Yews;
> The thriving plants ignoble broomsticks made,
> Now sweep those Alleys they were born to shade.[2]

Numerous critics have commented on the way the son's choices violate
various aesthetic rules and reveal his moral failings.[3] But by calling him
the "foe to . . . his *Father's* groves," Pope stresses that he rejects not merely
his father's superior aesthetic values but his *posterity*, in this case meaning
both the trees themselves and his father's sense of temporal connectedness. Given the cultural importance trees carried as living embodiments
of intergenerational connection, Pope's emphasis on the "thriving plants"
becoming "ignoble broomsticks" takes on special importance, as does the
fact that their demotion to "broomsticks" denies the trees the role they
were "born" to play. As living plants, trees not only connected present to
posterity but also provided "shade," a service to the public, making them a
central part of the socio-environmental community. Protecting and preserving that public service is as much a part of the duties of the good
landlord as protecting the trees for his posterity. The fact that Sabinus's
son rejects his father's useful and beautiful trees in favor of faddish landscape designs underscores his misplaced values. He literally and figuratively cuts off the continuity between his past, his present, and his future
and flouts his duties to his human and nonhuman publics. He loses both
the trees he inherited and his sense of his role as the steward of the gifts
of the past, the stability of the present, and the sustainability of the future. In his failure to fulfill his usufructuary duty to act as a guardian of
future good, he epitomizes the bad landlord. His failures, therefore, are
neither simply aesthetic nor symbolic. He has, quite literally, failed to
fulfill the obligations he is accountable for fulfilling as a usufructuary,
by failing to understand that there are limitations to his rightful use of
land—and power.

An idealized father-landlord contemplating his sense of intergenerational connection as embodied by trees; a son disregarding his
usufructuary duties to public and posterity by clear-cutting woods
to satisfy his short-term tastes: Pope's story of Sabinus and his son in
Burlington evinces strong topical and conceptual ties to the seventeenth-
century socio-environmental tradition of writing epitomized by Evelyn
and Finch. Concern with and admonitions about the getting and use
of "riches," both land and monetary, crop up likewise in the *Epistle to
Bathurst* and the *Horatian Imitations*. Taken together, this chapter will
show, the *Horatian Imitations* and the *Epistles* to Burlington and Bathurst
on the "use of riches" reveal in Pope's moral poetry of the 1730s a concentrated and extensive analysis of the problems of lordship, possession,
and the ethics of use that had been at the forefront of English culture
since at least the Restoration. Pope confronts the perennial questions of
English lordship: Under what moral conditions can unequal possession
be defended? And what, morally, does possession of and power over the
land everyone depends on require of the possessor?

In his sustained attention to the morality of possession and use in
Burlington, *Bathurst*, and his imitations of Horace's *Second Satire of the
Second Book* and *Second Epistle of the Second Book*, Pope distills and transforms the values of the usufructuary ethos he inherited to fit his cosmology and his politics. He also participates in a discussion about what the
contemporary challenges to those beliefs portended. The usufructuary
ethos provided a framework that could articulate the interconnections
of duty and dependence among God, humans, and nonhumans. Just as
importantly, it explained both the ethics of a hierarchically organized
universe and the limits of the power that hierarchy afforded classes of beings over each other. Pope's moral poetry confronts head-on the problem
that shadowed the writers that came before him: what happens when
someone with a great deal of power and property fails to fulfill their
duties—and, more disturbingly, fails to acknowledge that such duties
even exist, that possession entails obligation at all? What happens when
an entire culture begins to forget those duties? Under the usufructuary
ethos, the opposite of use is not reverence, but rather avarice, selfishness, and the breaking of the bonds of obligation and interdependency
between the self and others.

Pope's moral poems represent one especially prominent and influential
set of writings in a long series of literary and devotional expressions of
anxiety about possession and use that stretched from the late seventeenth

century to the late eighteenth, and intensified in both their anxiety and
their ambivalence as the transformative pressures of the emerging new
economy gained momentum. Pope was especially preoccupied with the
problem of avarice and what he saw as the emerging money economy's
tendency to amplify avaricious behavior. *Burlington* and *Bathurst* connect
money's capacity for abstraction to the abstraction of landscape itself.
Indeed, fundamental to the immoral uses of riches Pope decries are their
tendency to transform living environments into inert backdrops, severing the moral relationships among species and generations central to the
usufructuary ethos. His repeated poetic efforts to admonish proper use
by the wealthy landed class in his 1730s moral satires reflects both his
deep commitment to his version of the usufructuary ethos and his persistent anxiety about the fact that whatever the landed class *ought* to be
doing with their land, many of them were not.

The ways Pope's depictions of good and bad lords played into and
with Tory ideology—and took jabs at Walpole and his allies—have been
by now well canvassed.[4] Maynard Mack has also helpfully tied Pope's
1730s poems to the seventeenth-century country-house and retreat poetry traditions of Jonson, Marvell, Cowley, and others. Mack argues
that it is these poets' "feeling for the life of considerate use that pulses
behind much of [Pope's] social criticism," particularly in their common
focus on the virtuous life of the country gentleman in retreat from city
and politics[5]—at once the life Pope sincerely believed to be most con-
ducive to virtue, and a useful rhetorical position from which to critique
the moral decay of Walpole's London. At the root of that simultaneously
sincere and satirical stance, as Brean Hammond has shown, was Pope's
"ethical approach to political change," his "commitment to the ideal of virtue and to the possibility of ethical revolution" as a means to political and
social reform based upon the "belief that there are normative standards
of behavior as universal, simple, and available as the law of nature itself."[6]
Far from rejecting any of these important aspects of Pope's 1730s oeuvre,
I argue that refracting Pope's poems through the lens of the usufructuary
ethos clarifies the relationships among the various political, social, and
environmental facets of Pope's verses on the subject of the use of riches.
The usufructuary ethos, with its tenets of displacement, mediality, and
accountability, and its focus on benefit to public and posterity, provides
crucial ethical context for Pope's political and social arguments, grounding what in some lights appears to be purely partisan ideology in a hierarchical ontology with (as we have seen) deep roots in English philosophy,

theology, and literature. Those roots offer a philosophical grounding for Mack's observations of the poetic similarities between Pope and the prior century's retirement and country-house poems, but they also expand the literary and intellectual forebears of poems like *Burlington* and *Bathurst* beyond the tradition of complementary verse to patrons' estates, pointing to the deep confluences of political and environmental concerns in all such poems, particularly Pope's.

The *Horatian Imitations*, Usufruct, and the Jubilee Laws

In the course of his analysis of Pope's *The Second Epistle of the Second Book of Horace* (*Ep.* 2.2), Thomas E. Maresca points out an intriguing connection between Pope, the principle of usufructuary displacement, and a key verse from the so-called Jubilee laws of the Old Testament. Maresca writes:

> The only point at which [divine and earthly laws] have corresponded is in their mutual insistence on the evanescence of this world, a fact that renders earthly laws by their own decree absurd:
>
>> The Laws of God, as well as of the Land,
>> Abhor, a *Perpetuity* should stand. (246–47)
>
> The particular Law of God referred to in this case I take to be Leviticus 25:23:
>
>> The land shall not be sold forever: for the land is mine;
>> for ye are strangers and sojourners with me.
>
> The commentators explain this passage in a manner quite appropriate to Pope's present theme. God, they pointed out, is the Lord of all things, and man has use of them only for his lifetime: in effect, a capsule version of the poem's theme of God's *dominium* and man's temporary *usufructus*.[7]

Maresca ties Pope's invocation of Leviticus and his fixation on the ephemerality of property to the general Christian moral exhortation not to attach value to earthly things over eternal ones. Pope, he claims, is seeking "after enduring laws and those lands in which he will be more than a mere stranger and sojourner."[8] Maresca's primary interest in the *usufructus/dominium* distinction lies in the way he sees it shaping Pope's relationship to *poetry*, arguing that Pope is attempting to exemplify the "poet's proper assumption of *usufructus* and of the transmission of a cultural inheritance" through the use of tropes of law and property.[9] While that argument tantalizingly suggests the depth and extent to which the usufructuary ethos shaped Pope's poetry, it also diminishes the centrality of Pope's usufructuary ethics of actual use of the world in his *Imitations*, particularly his lingering preoccupation with questions of how the temporariness of possession affects the morality of the use of land and, by extension, with questions of the moral obligations of the English landlord.

Pope did not coin these connections. Leviticus 25 formed the basis for in-depth considerations of morality and justice of possession throughout the long eighteenth century, and commentators lingered on its usufructuary implications. According to Leviticus 25, the ancient Israelites were required to observe a "Sabbath year" every seventh year, completely resting the land, not sowing or tilling it, and living only off the accumulated produce saved up from previous years. Every seventh Sabbath year was to be a "Jubilee" year, when not only would the land rest but "ye shall return every man unto his possession, and ye shall return every man unto his family" (Lev. 25:3–5, 10).[10] All property that had been sold in the previous forty-nine years must be returned to its original owner in the Jubilee year; all slaves must be freed, all debts forgiven. The theological basis for this law was the belief that since the earth and all its creatures belonged permanently to God—the creatures were only "strangers and sojourners" on the earth—no human transactions could be permanent.

Early modern commentators on Leviticus focused on the ways that the Jubilee laws' insistence on humans' usufructuary relationship to creation encouraged just and equitable uses of land and power, and provided an example of an ideal against which contemporary systems of possession could be judged. Among the most widely read commentaries on Jewish beliefs and Old Testament law in the eighteenth century was the French lawyer and theologian Claude Fleury's *Les moeurs des Israelites* (*The Manners of the Israelites*), which presents the Jubilee laws as a bulwark against avarice and abuse of power, and as promoting liberty, labor, and just and equitable treatment of other stakeholders in the land, human and nonhuman alike.[11] Fleury takes special note of the way that the structure of usufructuary possession in the Jubilee laws protects against injustice and exploitation. Liberty, labor, and subsistence are linked in Fleury's analysis: to be free is to have full and secure control over the means of maintaining one's existence, implying that the land fundamentally ought to be understood as the source of survival, rather than of monetary wealth.[12] What makes the ancient Israelites an especially good example of that ideal, for Fleury, is the usufructuary nature of the Jubilee laws, which make the obligation to maintain the freedom and equity of society a part of the divine order of the cosmos itself. Fleury writes that the ancient Israelites

> could neither change Place, nor ruine themselves, nor grow too rich. The Law of the Jubile had provided against such like encounters, revoking every fifty years all such alienations, and annulling all obligations. By these means Disquiet and Ambition were retrenched: Every individual person applyed himself with affection to the improvement of his Inheritance, knowing, it would never go out of his Family. This Application was likewise a Religious Duty, founded upon the Law of God.... Moreover the Law says, that they were but the Usufructuaries of their Lands, God being the true Proprietor.[13]

Possession reverts to the original holders in Jubilee law on the premise that no permanent alienations can take place when humans have only the usufruct of the land, and Fleury argues that that basic premise establishes a particular relationship to the land—one that nullifies any tendency to transform land into a commodity to be accumulated for greater wealth and power, and that prompts the holder of the land to think of it instead in terms of the way it connects present, past, and future generations together through their continuing dependence on the land for their support. The usufructuary premise of the Jubilee laws structures human beings' relationship to the nonhuman world in a way that keeps their dependence (on it and on God) and their moral accountability (to others and to God) in view, linking justice in society and in distribution of resources to intergenerational and divine justice. The Jubilee redistributions may not be enforced by worldly sanctions, but they do tell the reader what uses of power and land are *right*.

These connections among land, time, and justice remained a theme in commentaries on the Jubilee laws throughout the eighteenth century, and later commentaries offered even more pointed rebukes to avarice and misuse among the wealthy, landed class. In his lengthy midcentury gloss of Leviticus 25:8, William Dodd emphasized the ethical and social implications of the Jubilee year, arguing that the Jubilee offered a

> curb to avarice; habituating [Israelites] to the exercises of humanity towards their slaves and beasts, of mercy and liberality to the poor: and Philo observes, as we have before remarked, that it was also a wise, political contrivance, to let the earth rest, in order to recruit its strength. . . . It provided against all ambitious designs of private persons, or persons in authority, against the public liberty: for no person, in any of the tribes, was allowed by this constitution to procure such estates as could give them hopes of success in oppressing their brethren or fellow subjects. . . . This equal and moderate provision for every person, wisely cut off the means of luxury, with the temptations to it from example.[14]

By emphasizing the limited and transitory nature of possession of God's creation, Dodd argues that the Jubilee laws not only enforce equality but also foster fairness in the *attitudes and behavior* of those subject to them. The constant reminder that everything is ultimately God's, Dodd asserts, makes kindness to dependents, fairness to fellow subjects, and steady, frugal labor not only the *right* kinds of behaviors but the most logical ones as well, increasing the likelihood that the Israelites would enact those behaviors of their own volition. The Jubilee encourages compliance with the moral laws under which God originally granted human beings the usufruct of the earth—laws which, Dodd implies, prohibit cruelty, tyranny, and greed and promote the equitable use and distribution of the means of survival. The Jubilee, in other words, functions as a guide to ethical behavior with regard to land use that is encoded into the usufructuary nature of human possession itself. Compliance with those laws, and with the spirit behind them, preserves the fair and just society that God designed.

Pope's hortatory responses to the culture and the behaviors of the landed class in his *Moral Epistles* and *Horatian Imitations* reflect precisely the same connections among displacement, justice, and morality of use that Jubilee commentaries featured. Pope repeatedly insists upon the irrelevance of the distinction between owning and renting based on humankind's fundamentally usufructuary position. In a mock dialogue with Swift in *Second Satire of the Second Book*, Pope writes:

"Pray heav'n it last! (cries Swift) as you go on;
"I wish to God this house had been your own:
"Pity! to build, without a son or wife:
"Why, you'll enjoy it only all your life."—
Well, if the Use be mine, can it concern one
Whether the Name belong to Pope or Vernon?
What's Property? Dear Swift! you see it alter . . .[15]

—after which he goes on to detail various ways that a person's property could be compromised or taken away. Similarly, in *Epistle* 2.2, Pope insists that use is the truest and, in fact, only type of ownership available to human beings. He writes that "*Use* can give / A *Property*, that's yours on which you live," and a few lines later derides the vanity of human wishes for permanent ownership:

The Laws of God, as well as of the Land,
Abhor, a Perpetuity should stand;
Estates have wings, and hang in Fortune's pow'r
Loose on the point of ev'ry wav'ring Hour;
Ready, by force, or of your own accord,
By sale, at least by death, to change their Lord.
Man? and for ever? Wretch! What wou'dst thou have?[16]

Maresca points out that in passages like these "Pope contends . . . that real ownership is impossible on earth, and that temporary use of objects is all that man can attain," an exact description of what I have called the displacement principle of the usufructuary ethos. Maresca points to the Levitical teaching of "God's ultimate ownership of all the objects man calls his own," as well as the "Roman legal distinction of *dominium* and *usufructus*" as background to Pope's argument.[17] Aquinas, Maresca notes, was the main theological precedent for the Christian version of this concept, but as Fleury and Dodd show (not to mention Allestree, Hale, and other devotional writers studied in previous chapters), Pope would have been likely to have encountered the concepts of the usufructuary ethos regularly in a variety of texts throughout his life and career. On the classical side, Maresca notes that when "Pope describes himself as a tenant rather than an owner, or a user rather than a possessor" of his land, it "is in part drawn from the famous dictum of Lucretius: 'For life is not confin'd to him or thee; / 'Tis giv'n to all for use; to none for Property,'" the latter quote coming from Dryden's translation of the famous, and highly usufructuary, passage of Lucretius, with which Pope would undoubtedly have been familiar both in Latin and in translation.[18]

Pope not only returns again and again to the displacement principle but also, echoing Evelyn and Finch in the previous chapter, returns again and again to the usufructuary possessor's accountability to the public and posterity. In *Satire* 2.2's chiding of the "impudence of wealth! with all thy store, / How dar'st thou let one worthy man be poor!"[19] for instance, Frank Stack detects an "emphasis on public duty . . . [that] suggests the freshness of [Pope's] response to the Latin"; Andre Dacier's French commentary on Horace's *Satires*, Stack notes, "does not stress this at all."[20] Pope links the duty to the public with the duty to posterity in *Satire* 2.2, suggesting that transgressions against others in the present and others in the future are connected by the fundamental mistake of assuming that "your" possessions are yours, rather than being temporarily entrusted to you for the betterment of all. In one of his most cutting passages, Pope writes:

> When Luxury has licked up all thy pelf,
> Curs'd by thy neighbors, thy Trustees, thy self,
> To friends, to fortune, to mankind a shame,
> Think how Posterity will treat thy name;
> And buy a Rope, that future times may tell
> Thou hast at least bestow'd one penny well.[21]

The list of those who "curse" the profligate lord are also those to whom he was accountable for the use of his gifts: "neighbors," "trustees," "friends," "mankind," "Posterity." This is the web of people whose lives are linked through the choices the individual lord makes about the use of "his" possessions, whom he can affect for better or ill. When combined with Pope's insistence upon the extent of true human "ownership" being limited to use, his savage denunciation of misuse by someone society and law would recognize as an "owner" indicates the ethical centrality of the concept of "use" to Pope's 1730s oeuvre. What human law might allow an owner to do with his property has no bearing on his ethical rights or duties as dictated by his ontological status as a mere usufructuary. What you choose to do with your "riches" during your lifetime, and whether those choices fulfill the duties associated with your status—that is what matters. All this indicates the unique importance duty to the public and posterity held in Pope's conception of the ethics of the use of riches, including the use of land. The biblical commentaries dealing with the nature of human possession were saturated with discussions of justice and duty, all pointing to the moral necessity of understanding one's self and possession as temporary and transitory, and expressing suspicion and anxiety about framing land as a way to accumulate wealth or power.

TIME, "SENSE," AND "MAGNIFICENCE" IN *EPISTLE TO BURLINGTON*

Pope's most in-depth development of the issues of the usufructuary ethics of use of the nonhuman world appears in his *Epistle to Burlington*. The poem contrasts proper uses of land—specifically those that uphold the chain of posterity—with uses that sever the connections among beings and generations by focusing on gratifying only the appetites of the present moment. He distinguishes the two types of use by the concepts of "sense" and "taste." "Sense," this section will show, is defined by the same criteria that defined good usufructuary landlordship. "Good Sense, which only is the gift of Heav'n," is a sensitivity to the features, functions, and needs of the land and the lives of the beings that live and depend on it, and more particularly, to their interconnections with each other across space and time.[22] The example of Sabinus at the beginning of this chapter provided the first example of the ethical importance *Burlington* places on sensing one's connectedness to place, public, and posterity. What we find in this poem is a merging of the values of the usufructuary ethos, Pope's aesthetic principles, and his ontology of the interdependent Chain of Being into a vision of the socio-environmental landlord that has its roots in, but is distinct from, that of earlier writers like Evelyn and Finch.

The majority of critical attention to *Burlington* has focused on Pope's particular brand of landscape and architectural aesthetics. Insofar as the ethics of use have come into discussion, scholars have mostly focused on the ways various characters in the poem do or do not adhere to classical standards of architectural decorum, or the ways Pope's admonishment to "Consult the Genius of the Place" functions as a salvo in his establishment of his own landscape aesthetic against the older French style.[23] *Burlington* culminates in (to quote Pope's "Argument") a discussion of the "proper Objects of Magnificence, and a proper field for the Expence of Great Men"[24]—that is, a statement on the ethics of the "use of riches"—which Earl Wasserman famously attributed to an exclusive focus on

Aristotelian aesthetics and ethics.[25] More recently, William A. Gibson similarly claimed that "'Use' and decorum are architectural principles; 'Magnificence' is an effect in building that can be achieved by adhering to 'Use' and decorum," attributing all the poem's discussions of proper and improper use of riches to issues of good and bad architectural choices.[26] James Engell has argued that, for Pope, "virtue [is] what we do with . . . the semiotic systems at hand," both monetary and literary, and "how we *bestow* value through motive and acts."[27] Others have claimed that Pope appeals to the cultural semiotics of nostalgic aristocratic apologetics as a way to defend sociopolitical or economic affiliations with various members of the landowning classes.[28] In all these cases, the nonhuman world of *Burlington* and Pope's interest in it is presumed to be primarily, if not exclusively, symbolic of something other than itself.

The limitation of such scholarly perspectives is that they presume perhaps too strictly that the Augustan ethics of use applied exclusively to human beings and their sociopolitical constructions, to the exclusion of humans' ethical obligations to that which was *being used*. But that was demonstrably not the case. "Magnificence," for example, had a long-established definition in Christian ethics that had nothing to do with architecture or aesthetics. In the *Parson's Tale*, Chaucer wrote that "magnificence" is "whan a man dooth and perfourneth grete werkes of goodnesse; and that is the ende why that men sholde do goode werkes, for in the acomplissynge of grete goode werkes lith the grete gerdoun."[29] Christian "magnificence" is about "goodness," that is, doing works that do good *for others*. As Aubrey L. Williams has observed, at the heart of *Burlington* is a moral lesson about the necessity of *charity*, a Christian concept deeply connected to both "magnificence" and "use." In his reading of *Burlington*, Williams connects the sort of "use" the poem promulgates to the "traditional specifically Christian tenets . . . which admonish that the rich man is required . . . to act as God's earthly steward,"[30] and he quotes as support seventeenth-century theologians espousing the usufructuary view of human possession in language that is, by now, familiar to us.[31] Williams points out that the basic moral inherent in such devotional writings matches that espoused by *Burlington*, that "Use alone . . . sanctifies Expence," and specifically the sort of "use" that fulfills the duties possessors owe to public and posterity. But while Williams restored to prominence the ethical depth of the poem that had been denied by its previous readers, his anthropocentric assumption that the poem's moral framework was confined to human beings caused him to overlook crucial aspects of both the poem and his theological sources that point to the centrality of the nonhuman world to Pope's ethical vision. In Pope as in the writers before him, the necessity of charity and doing "great works of goodness" is shared among all beings and across time.

Thus, when the lines describing the "proper objects of Magnificence" begin with Pope's claim that "'Tis Use alone that sanctifies Expence, / And Splendor borrows all her rays from Sense," he refers quite clearly to the parameters of "use" adumbrated in the previous chapter, requiring (in Swift's roughly contemporaneous words) that all "Works of Nature . . . be useful, and in some Manner a Support to each other." And he links that definition of "use" to his own concept of "sense," that indispensable sensitivity on the part of the landlord to his place in the temporal chain linking his ancestors with his descendants:

> His Father's Acres who enjoys in peace,
> Or makes his Neighbors glad, if he encrease;
> Whose cheerful Tenants bless their yearly toil,
> Yet to their Lord owe more than to the soil;
> Whose ample Lawns are not asham'd to feed
> The milky heifer and deserving steed;
> Whose rising Forests, not for pride or show,
> But future Buildings, future Navies grow:
> Let his plantation stretch from down to down,
> First shade a Country, and then raise a Town.[32]

Pope's epithet for the cow—"milky heifer"—draws attention to the way the cow fulfills her obligation of "mutual support"; she produces her offering of milk, fulfilling her providentially designed role in providing sustenance for other creatures, and humans must respond in kind, using their gifts and abilities to support the cow. The "steed," likewise, is by virtue of its own role in aiding the lives and labor of others on the estate "deserving" of care and support—and deserving in the sense of being rightfully owed that support. Properly managed, the lord's property sustains the health and happiness of both its human and nonhuman dependents, whose well-being in turn supports the landlord's health and happiness. Proper use, according to *Burlington*, not only produces benefits for landlord, tenant, and neighbor but also recognizes and supports the labor of nonhuman tenants as well.

Furthermore, proper use guarantees the continuation of a tradition of health and happiness into the future. These are, after all, his "Father's Acres," which, along with sustaining tenants and cows in the present, will provide for their posterity. Use directed by sense requires attention and deference to the needs of fellow beings (human and nonhuman) whom the land has traditionally supported, as well as to the needs of those who will rely upon it in the future. The parallel with the usufructuary concern for intergenerational justice is obvious: by taking the estate's past and future functions into consideration in plans for the present, sense helps guarantee the well-being of future generations. Following the Sabinus story earlier in the poem, Pope once again makes the framing of property as a material link to past generations of central importance in defining the sort of landlord who shall "grace" and "improve the Soil."[33] The final line of the verse paragraph completes the temporal bridge from past to future, insisting not just on the importance of trees' past as the "Father's Acres" or their roles as "future Buildings, future Navies" but on the importance of every stage and moment of their lives: "*First* shade a Country," Pope writes of the "rising Forests," and "*then* raise a Town." The trees' past and present are as important as their future, and Pope's final line, reminding his reader that the trees that have become the boards in a house or ship were once an equally important living part of a rural community, negates the possibility that trees could rightly be seen as mere standing reserve to a landlord of sense. More importantly, we see once again in this passage the interconnection of public and posterity in the usufructuary ethos, as Pope subtly reinforces that the public—in this case, national—benefit is dependent upon conscious concern for and attention to duties to posterity, and to the awareness of one's self as the previous generation's posterity and of the existence of one's own posterity to come.

"Sense" and Sustainability

Connectedness and continuity are the hallmarks of sense in *Burlington*. Its ethical importance derives from the way that possession of sense enables the landlord to perceive and preserve his estate as the living embodiment of a community that crosses the boundaries of species and time. This vision of the thick interdependence among living things echoes the ontology Pope laid out in the *Essay on Man*. *Burlington*'s command to "Still follow Sense, of ev'ry Art the Soul, / Parts answ'ring parts shall slide into a whole" may in some contexts seem to exhort a merely aesthetic unity,[34] but juxtaposed with *Essay on Man* the idea of "parts" becoming a "whole" takes on deeper significance:

> Nothing is foreign: Parts relate to whole;
> One all-extending, all-preserving Soul
> Connects each being, greatest with the least;
> Made Beast in aid of Man, and Man of Beast.[35]

In Pope's ontology, the universe is inherently both hierarchical and inter-dependent. Everything, regardless of its place in the Chain of Being, is "related" to every other thing, and every thing is created "in aid of" every-thing else. The source of that connection, moreover, is an "all-extending, all-preserving *Soul*" that binds together every being and action. In *Burlington*, that force takes the form of "Sense," and it does the very same work of enabling "parts" to "slide into a whole" within the microcosm of the estate.

Crucially, the microcosm of the estate is directly analogous to the mac-rocosm of the universe; every being that lived, lives, or will live in that microcosm is linked with every other for their collective well-being. Un-derstanding that analogy makes it possible for the landlord in possession of sense to "read" the proper order of his estate and his place in it in the natural world itself. When Pope commands the reader of *Essay on Man* to "Look around our World" and "behold the chain of Love / Combin-ing all below and all above," he implies that natural law is immanent in the natural world, and that observing nature can teach humans their di-vinely appointed "relation and condition" and its attendant moral duties.[36] Courtney Weiss Smith has shown that, although apparently politically and philosophically diverse, Pope and Locke both "find . . . in nature" in-structions for proper natural and social relations, depicting humans first learning about the plan of creation from nature, then patterning their own behaviors and societies on it.[37] Meditating on and drawing moral, religious, and political conclusions from nature was characteristic of late seventeenth- and early eighteenth-century English writers. Smith's obser-vation recalls the way Allestree and other theological writers drew from nature's example to exhort humans to fulfill their divinely ordained obli-gations for the benefit of the rest of creation. Allestree remarks, "The very *inanimate* Creatures afford their consort to this divine harmony; every one of them perform those offices, fail not in the exercise of all those (not unactive) qualities God hath put into them. . . . all this . . . is most visible without help of a perspective."³⁸ To "follow Nature," for Pope and his con-temporaries, was therefore never merely a question of proper aesthetics, but rather one of reading the intentions and injunctions of the creator through the example of his creation. Sabinus's moral superiority in *Burl-ington* derives from his ability to understand correctly his relationship to and role in the microcosm of the world that is his estate—as a usufructu-ary lord, accountable to his dependents and descendants. That is the truth he is able to "read" in his "young Woods," the truth his son fatally ignores.

Hence it is that Pope insists that the act of planning or planting or building alone is insufficient to create an ideal estate. Its preservation through time is of equal importance, and sense includes the willingness to let the place mature—let its living beings grow—without untimely inter-vention. There are two factors necessary to making one's estate a "work to wonder at," Pope writes—the participation of "Nature," and the passage of time: "Nature shall join you, Time shall make it grow."³⁹ Nature "joins" in the sense of both cooperating with the landlord's efforts providing that he is following the "genius of the place," and in the more active sense of "joining" the "parts" so that, as Pope writes a few lines above, they "slide into a whole."⁴⁰ But cooperation with nature goes beyond making the right kinds of landscaping plans. "Time" is the other necessary ingredi-ent. A truly magnificent estate, in the full Christian sense, is the product not merely of proper planning but of sustainability through time, of its ability to persist and thrive, and of the landlord's willingness to leave it alone and let it grow. Its "greatness" is organic, not built, the outgrowth of the inherent properties of the living things that constitute it, not of the landlord's will or aesthetic sense alone. Sense, in *Burlington*, thus refers to a sensitivity to the living connections that each choice can foster or break. It refers not only to the aesthetic "sense" but to the possession of a sense of what your place and duties in the world are, and the sense to fulfill them, and the sense to prioritize the long term over the short. To possess sense is to understand oneself in the context of relationships that stretch across both time and space.

The subsequent lines illustrate the necessity of time for growing an estate reflective of proper sense. Pope illustrates his principles through a series of contrasting examples: one example of a man of with proper tem-poral sense, Sabinus, sandwiched between two examples of men without it, starting with "Villario." Villario, who has spent a decade planting his grounds, is at last seeing his work come to fruition:

Behold Villario's ten-years toil compleat;
His Quincunx darkens, his Espaliers meet,
The Wood supports the Plain, the parts unite,
And strength of Shade contends with strength of Light.....[41]

The trees Villario has planted are on the cusp of reaching both aesthetic and biological maturity, individual specimen parts becoming the harmonized whole of the estate. The still-growing branches have begun to provide that ideal Popean balance of "Shade" and "Light," a "World harmoniously confus'd."[42] The different "parts" of the landscape "unite" to form this new whole, as befits a landscape that has "followed Sense," and has allowed nature to "join" these disparate pieces and time to let them "grow." What's more, there is a sense of cooperation in Pope's choice of verbs in this passage: the "Espaliers meet" one another, as their natural growth guarantees they will, while the "Woods support the Plain"—their differences not conflicting with one another, but rather offering their own unique, but mutually beneficial, gifts. And while Villario has used his unique human gifts to plant these trees and to nurture their growth, the trees now offer up their own gifts, as they were designed to, of shade, and of fruit.[43] The harmony Villario has achieved is both aesthetic and environmental; the pleasing nature of the space comes both from its attractive, pleasant qualities and, more importantly, from the way in which the design of the landscape makes it possible for each living thing to fulfill its role of being of beneficial use to each other living thing. And, most importantly, those benefits promise to be extensive and lasting. Shade and fruit and human husbandry are not onetime gifts, nor are they gifts that benefit only one person. This is a landscape that could benefit both the public and posterity.

Still, Villario is ultimately an example of a landlord who *lacks* proper sense. His failure is not in his design, which as we have seen fulfills Pope's aesthetic and ethical standards. Rather, Villario lacks a proper sense of *time*, focusing selfishly and shortsightedly on his aesthetic enjoyment. Just as his trees are growing into their roles, Villario impatiently decides to cut them down: "Tir'd of the scene Parterres and Fountains yield, / He finds at last he better likes a Field."[44] The heart of Villario's moral failure lies in the way that his decision reveals his insensitivity to the very qualities the previous lines had emphasized: the ways that his trees embody the cooperative energies and mutual gifts of human beings, plants, soil, and sun to create a landscape that balances the needs and benefits of all members of that multispecies community. That Villario tires of the *"scene"* reveals that he has, all along, reduced the living community of beings he created to props whose sole purpose was to satisfy his aesthetic impulses. Or, to put it in a more *Burlington*-ian way, he has all along seen his estate merely as a means to satisfy the transitory demands of his *taste*. His choices are motivated by what he likes, without any sense of his estate as a place that exists to support a community of living things, not only in the present but into the future. Villario can undo the work of a decade without compunction because he lacks a sense of communal accountability, and of the landscape as existing beyond himself, as a living connection to past and future. Villario's decision is wasteful, and his rejection of the "Parterres" for "Fields" aesthetically dubious in Popean terms, but those aesthetic failures reflect his fundamental moral failure to understand the basis for a proper "use of riches."

The sort of "Use . . . that sanctifies Expence" in *Burlington* follows the usufructuary principle that all creatures share an obligation to benefit the lives of others, an approach to the management and use of property that actively and consciously pursues the good of both public and posterity. Villario's negative example returns us to where we began this chapter, with Sabinus, whose story immediately follows. Placed in the context of the specific nature of Villario's failure, Sabinus's pleasure in watching his "young Woods" as they grow, "stretching" as they "long to meet," just as Villario's had, takes on additional significance. By juxtaposing Villario and Sabinus, Pope underscores that the key difference between them is that Sabinus derives his pleasure from observing the living trees grow into their places in the landscape, and from his anticipation of the comfort their long lives will provide to future generations. Unlike Villario, Sabinus displays a correct sense of himself and his duties as a landlord, and expresses that understanding through his actions, cooperatively fostering an environment that will sustain and benefit the creatures that will be there after he is gone. His aesthetic enjoyment is bound up with his awareness of what the trees will contribute to public and posterity. True aesthetic value, for Sabinus and throughout *Burlington*, is inextricable from ethical value.

Thus, in light of the usufructuary ethos, Pope's theories of "following Nature" and "Sense" in *Burlington* indicate that a belief in the inherent value of nonhuman nature underlies his deep moral concern with the use of the land. Engell rightly identifies Pope's "Sense" as the "virtue of human action, the benefit or damage caused by motives directing the

use of words and wealth," but he presumes that value is human beings'
to *bestow* on nature through their use, rather than an inherent quality
that they honor *through* proper use.[45] For Pope, nature is not inert, or
a mere backdrop to human activity, or devoid of its own moral significance. Land is not simply material upon which human beings project
their vision, nor are the aesthetic qualities of "taste" the locus of the
moral importance of the issues of land use in Pope's moral epistles.
The "linkage of *taste* with virtue" may be a "non sequitur," as Malcolm
Kelsall as claimed, but the linkage of *sense* with virtue is not, because it
reflects in the use of the land an awareness of the interdependency of
the hierarchy of beings, and the deeper ontological and moral significance of whether or not one fulfills one's duties and enables other parts
of creation to fulfill theirs.[46]

EPISTLE TO BATHURST'S "MAN OF ROSS" AND THE USUFRUCTUARY LANDLORD

Maynard Mack has written that the *Epistle to Bathurst* asks whether
"riches, or any other good" are "a gift from heaven, as in the parable of the
talents, or our own 'lucky hit'? Does the accessibility of the earth's beauty
and bounty to our use call for ownership or stewardship?"[47] Pope's answer to that question is, quite evidently, the latter. But the centrality of
Bathurst's endorsement of the usufructuary ethos's insistence on displacement, mediality, and accountability, as well as the poem's continuity
with the landlordly values of *Burlington*, has tended to be overlooked by
previous scholars in their (understandable) preoccupation with *Burlington's* neoclassical aesthetics and *Bathurst's* political critiques. If we read
Bathurst in dialogue with *Burlington* and Pope's *Horatian Imitations* of
the same period, however, it quickly becomes apparent that in all these
poems Pope espouses a single vision of a proper "use of riches" that is
in line with the usufructuary ethos, in particular with those facets of it
that emphasize human beings' obligation, owed in common with every
other part of creation, to share their bounty and gifts for the support
of other creatures. The key to this aspect of Pope's thought in *Bathurst*
comes in the middle of the poem, just before his introduction of the exemplary Man of Ross:

> To Want or Worth well-weigh'd, be Bounty giv'n,
> And ease, or emulate, the care of Heav'n,

Whose measure full o'erflows on human race;
Mend Fortune's fault, and justify her grace.
Wealth in the gross is death, but life diffus'd,
As Poison heals, in just proportion us'd. . . .[48]

The first two lines of this passage are particularly complex, with "the care of Heav'n" operating on two distinct and equally important levels. In the first, "care of Heav'n" is grammatically a noun, the object of the verb "ease": "the care of Heav'n" are those creatures whose "want" must be "eased" by being "giv'n" the "Bounty" of the divine, either through direct access to God's gifts in nature or through the sharing of "Bounty" by human beings who have received more riches than others. In the second level of meaning, "care" operates as a verb: those to whom "Bounty" has been "giv'n" prove their "Worth" by "emulating" the "care of Heav'n," meaning those acts of the divine natural order that provide care and support for the rest of creation. This is primarily done, as the rest of the passage goes on to imply, by literally sharing the wealth. As Pope's lifelong close friend Jonathan Swift wrote in his sermon "Of Mutual Subjection," "God hath contrived all the Works of Nature to be useful, and in some Manner a Support to each other . . . we are obliged to act, as far as our Power reacheth, towards the Good of the whole Community."[49] Bathurst follows Swift and other contemporary devotional writers who insist upon the moral obligation to share the gifts of riches, material and otherwise, *and* in pointing out that that obligation is one shared in common by all beings, observable in every aspect of nature. The noun and verb senses of "care" converge here: the rich man must "emulate" both the divine act of caring for creation and the way that other creatures, aka the "care of Heav'n," themselves pour their gifts out for the benefit of others. In pointing simultaneously up the chain of being toward God and down toward other earthly creatures for examples of the unifying, universal obligation to share "riches" for the benefit of all, Pope recalls popular seventeenth-century writers like Adams and Allestree who preached the same idea, indicating the influence of the values of the usufructuary ethos on Pope's moral poetry.

Pope amplifies the duty of care in the *Second Satire of the Second Book of Horace, Paraphrased*. In response to an imaginary lord claiming the right to "have a taste" (note the overlapping Popean vocabulary with *Burlington's* "Visto! have a Taste")[50] because "'in me 'tis noble, suits my birth and state, / 'My wealth unwieldy, and my heap too great,'" Pope chides:

Then like the Sun, let Bounty spread her ray,
And shine that Superfluity away.
Oh Impudence of Wealth! with all thy store,
How dar'st thou let one worthy man be poor?[51]

In light of the devotional trope that underscored the ontological and
moral parallels between the rich man and various aspects of the rest of
creation, Pope's "like the Sun" takes on a significance that transcends
metaphor. As Thomas Adams had argued in a sermon a century earlier,
since the "earth, in a thankefull imitation of the Heauens, lockes not vp
her treasures within her owne Coffers," the rich man ought to be able to
perceive his role as a the conduit of resources through mere observation
of the natural world.[52] This sort of "analogizing" was, as we have seen, a
characteristic mode of synthesizing and connecting theological, empirical, moral, political, and social ideas for Pope and his predecessors, and
of explaining the ways the natural order was interlaced with theological,
social, and political truth.[53] What's more, Pope's simile is not present in
Horace's original Latin. The lines that follow decrying the "Impudence of
Wealth" paraphrase Horace, but the admonishment to imitate the sun's
rays has no equivalent in the original poem: the image is entirely Pope's
own.[54] By interpolating that couplet, Pope invokes and participates in
the discourse of the usufructuary ethos, which imbued such similes with
a literalness that underpinned their moral weight. The call to imitate the
sun is no pretty conceit; rather, it is a reminder to the reader that the divine design of the universe rests upon the mutual obligation of all its
denizens, inanimate and animate, to carry out God's design, a design
that relies upon the sharing of God's property for collective support and
survival.

Burlington's Sabinus and Villario showed that a successful landscape
for Pope requires the cooperation of human beings, nonhuman beings,
geography, and time. This is equally true in the Epistle to Bathurst. The
conjunction rests in the two Epistles' shared subtitle: "Of the Use of
Riches." In Bathurst, the ideal exemplar of the "use of riches," the Man
of Ross, "emulate[s] the care of Heav'n," using his estate in ways that "follow Nature" by supporting the well-being of his fellow creatures. The
Man of Ross passage begins with a series of rhetorical questions:

Who hung with woods yon mountain's sultry brow?
From the dry rock who bade the waters flow?

Not to the skies in useless columns tost,
Or in proud falls magnificently lost,
But clear and artless, pouring thro' the plain
Health to the sick, and solace to the swain.[55]

What I would like to draw attention to in these lines is the purpose-
ful blurring of agency. Pope does eventually answer these questions by
placing the Man of Ross's name in the mouth of "each lisping babe."[56]
But by suspending resolution of these questions for half a dozen lines,
Pope prompts his reader to consider the full range of possibilities the
questions open up, pointing to a more complex and interconnected set of
agencies. The Man of Ross chose to plant those trees on the "mountain's
sultry brow" and decided not to build fountains or other artificial water
features. But as *Burlington* taught us, deciding to plant trees is only the
first step; the trees themselves must grow, and the planter must have
the patience and foresight to allow them to attain their full maturity and
fulfill their greater natural and public purpose in order for the choice to
redound to his moral advantage. The "woods" on the "mountain's sultry
brow" are, therefore, indicators not merely of the Man of Ross's goodness
or skill alone but also of his cooperation with nature and time, and his
proper sense of his social, environmental, and temporal place and role.
Pope's goal in recounting this passage is to hold up the Man of Ross as a
moral exemplar for others to follow, and he achieves this through praise,
but a fundamental piece of that praise emphasizes the cooperative and
collective aspects of the Man of Ross's stewardship; his awareness, like
Sabinus in *Burlington*, that his job is to facilitate the passage of "riches"
socially and temporally to those that need them.

The line asking "who bade the waters flow" from "the dry rock" underscores the collective and cooperative nature of the Man of Ross's
work for the public benefit. Phrasing the question in that specific way
invokes the story of Moses in the desert, who struck a rock to provide
water to his parched people. But the proper answer to the question of
who "bade the waters flow" from the "dry rock" in Moses's case is actually
God, who made it possible for the rock to pour forth water when struck
by Moses. Insofar as the Man of Ross had the *power* to use water in
wasteful, fashionable ways and chose not to, he ought to be praised. But
Pope's phrasing of the question reminds the reader that it is ultimately
God who supplies all "riches," including water, and that "pouring thro'
the plain / Health and solace" was, after all, the purpose for which God

designed water to flow through an area in the first place. Thus the Man of Ross does not himself *cause* the water to flow nor give it its power of providing health and solace. What he does is make proper use of the gifts God made available to him.

By making it possible for his land and water to fulfill their roles of providing shelter and refreshment to other creatures, the Man of Ross fulfills his own role as an ideal lord. His "Cause-way parts the vale with shady rows," indicating that trees along the Man of Ross's roads fulfill the socio-environmental role Evelyn had described for them in *Silva*, providing protection from the sun for travelers.[57] This is what Pope means when he writes that the "due Medium, and true use of Riches" is not only the "*sense* to value" them and the "art / T'enjoy" them, but also the "the virtue to *impart*" them.[58] As in *Burlington*, the "sense to value Riches" properly requires understanding that they are not yours to keep. (That usufructuary possession also balances the right to "enjoy" something temporarily before eventually imparting it to someone else is relevant here.) The Man of Ross embodies the ideal landlord because he honors that balance, prioritizing his social and ontological duties to "impart" his "riches," and deriving enjoyment from the fulfillment of that duty. He has a proper sense of his own temporal and ontological position as a temporary and accountable usufructuary of the world, and of having a duty to organize things such that his "riches" pass through possession to those now and in the future who will rely on them, as the water of his streams passes through his land to pour on the plains and swains. The Man of Ross is ideal, therefore, because he embodies the usufructuary values also adumbrated in the *Epistle to Burlington*.

The Misuse of Riches

One curious fact about Leviticus 25's much-vaunted Jubilee laws: they were almost certainly never enacted, even in ancient Israel. The biblical scholar Jeffrey Fager notes that ancient Jewish priests' preoccupation with the Jubilee laws was likely a response to the fact that the sorts of ecological, labor, and economic exploitations the Jubilee laws decried were actually taking place, and the priests felt that some kind of divine moral decree was necessary in order to stave off injustices that were tolerated by the legal and economic system.[59] The situation in early eighteenth-century England was, arguably, similar. As previous chapters have shown, the consistent and insistent appeals to usufructuary values in devotional and literary works speak more to writers' observing those around them deviating from those values than to their unquestioned authority. Recall, for instance, the impetus for Evelyn's *Silva*: the decimation of England's trees in service of various parties' pursuit of short-term profit. Finch's "To My Lord Winchilsea," similarly, was written in response to the actual destruction (and eventual restoration) of the landscape; and Philips dedicated his adaptation of the *Georgics* in part to grappling with the moral challenges of contemporary land use.

Pope, therefore, was far from the only writer of the early decades of the eighteenth century to be preoccupied with the question of what those who had greater "gifts" of property and power ought and ought not to do with them. Anxiety about the problem of misuse of riches, whether through avaricious accumulation or prodigal waste, was a common theme in writing of the first half of the century across genres. This discourse had roots in traditional exhortations of moral behavior for the rich and powerful, but it took on special intensity in response to the financial and commercial transformations that accelerated over the course of Pope's life, particularly as Walpole's regime deepened the political rift between the so-called land and money interests. As the avenues and opportunities for profit widened, both for owners of English land and through investment in colonialist projects and other financial instruments, it became necessary to make finer distinctions between what kinds of "gain" were acceptable or unacceptable. It also became necessary to articulate the socio-environmental duties landowners ought to be fulfilling, and to delineate the pitfalls of failing to do so, in more stringent and detailed ways. Always at the heart of these texts is an awareness of the problem of power. Belief in and adherence to usufructuary duties can be urged, chided, commended—but it cannot be compelled.[60]

Pope was well aware that that the potential for abuse of property, whether of the avaricious or prodigal variety, always existed. But like others, such as his friend Bolingbroke, he believed that the social and cultural changes associated with the financial revolution made such abuse easier and more attractive to perpetrate. Jacob Fuchs locates Pope's sense of the new threat that developments of the English economy posed to human-nature relationships in one of the disparities between Pope's *Satire 2.2* and Horace's original. In contrast to Horace, in *Satire 2.2*,

ownership [of land] continues, but use, which Pope celebrates elsewhere, most clearly in the *Epistle to Burlington*, seems to stop. As a result perhaps, in an odd echo of the magic within Twickenham, the land outside of it seems mysteriously to change its character. It becomes a "Lawyer's share" or a consideration in a "Case"; it can even fall out of existence altogether: "... vanish from the Heir [or air?]." Since all of this unwinds next to the Latin poem's vision of land eternally yielding to use by the race of Ofellus, the contrast makes land in England seem literally denatured, fatally altered by man's blindness to what it really is.[61]

The fundamental source of the threat that money posed to the usufructuary ethos, in Pope's eyes, is the fact that financial and legal developments have the power to make landowners blind to "what it really is," and therefore to what they owe to it, to its creator, and to other beings that live on it. Pope contrasts what *is* with what *seems* again and again, tying the proper use of land to the former and monetary wealth to the latter. Pope insists he was in "South-sea days not happier, when *surmis'd* / The Lord of thousands" than in "five acres now of rented land."[62] In neither case did Pope actually possess the property he's associated with; to be Lord of "thousands" that exist only theoretically and can and did vanish is to be less a lord, Pope implies, than one is as a renter. The main difference is, as a renter, Pope has the *use* of the land, and the ability to fulfill all the most important and meaningful aspects of possession: "if the Use be mine, can it concern one / Whether the Name belong to Pope or Vernon?"[63] Since possession is always displaced and usufructuary anyway, the use is what's important—and proper use is precisely what things like stocks and mortgages undermine.

Much work has been done to try to pin down Pope's exact attitude toward money, paper credit, and the individuals associated with them.[64] My primary interest is not in Pope's stances on particular people associated with the political debates of his day (which in any case is, as the scholarship of the last twenty-five years has shown, a complicated and difficult question to answer).[65] Rather, I am interested in Pope's diagnosis of the kinds of misuse and abdication of duty that the rise of economic abstractions like paper money or stocks made particularly easy and accessible. As James Engell has argued of the *Epistle to Bathurst*, "Pope recognizes how money in the form of the written word can be misused with particular ease and stealth. Intent to keep an eye peeled for the end or uses to which wealth is put, whether in goods, gold, or paper, he is painfully aware how paper especially—that is, writing—offers opportunity for an empire *without bound* of deception and avarice."[66] The fact that the new forms of wealth associated with paper and writing make possible at least the illusion of an "empire without bound" sends up warning signals in the context of the values of the usufructuary ethos that Pope subscribed to. After all, the ethics of usufruct depend upon explicit boundaries and limitations to use and possession. The problem Pope saw, as Mack puts it, was that the "current ruling class . . . was abdicating its cultural responsibilities and losing ground as a political and social force" as a result, and that problem was accelerating owing to a "rising class that much too often acknowledged no responsibilities at all."[67] Specifically, the "responsibilities" they failed to acknowledge were their responsibilities to their socio-environmental communities, present and future, as usufructuary lords. The core of the issue in *Burlington* and *Bathurst*, the last parts of this chapter will argue, comes down to a problem of perception: the false attractions of "taste," luxury, and greed prevent the misusers of Pope's *Epistles* from seeing and fulfilling their duties to others and to the future. Their misuse has profound consequences both for themselves and for their communities, and, ultimately, for the usufructuary ethos itself.

EPISTLE TO BURLINGTON, "TASTE," AND THE PROBLEM OF THE PRODIGAL

The most important and the most critically overlooked of the mistakes the improvident lords of the *Epistle to Burlington* share is that in their pursuit of "taste," they treat their land as if it were populated by *objects*, rather than by *living beings*. Living beings are meant to be of "use," of course, but of use in that special usufructuary sense of collective usefulness—the sense of existing *for the good of* other living beings, of performing works of "goodness" for others. Not only is Sabinus's son the "Foe" to "this Father's groves," highlighting his failure to preserve his posterity and the embodied connection between past and future, but Pope's repeated linguistic juxtapositions of images of life and death underscore the fact that Sabinus's son has replaced a living landscape with a dead one. The "thriving plants" were "ignoble broomsticks made," and they now "sweep those Alleys they were born to shade."[68] In repudiating his own birthright by cutting down the trees, Sabinus's son has denied the trees their birthrights as well. Since offering shade is among the most important public benefits Evelyn assigns to trees in *Silva*, the living trees' demotion to dead broomsticks points toward Sabinus's son working against the universal order of creation, as well as against the trees and his father. Crucially, the source of that mistake is the son's "fine Taste," which desires an "op'ner Vista."[69] Taste, a myopic, anthropocentric worldview, sees the nonhuman world as a backdrop and a repository of materials to express the human landlord's aesthetic visions. It turns the landlord's attention away from his place in the socio-environmental order as a whole and toward the expectations of fashion. The fact that the *Epistle to Burlington* was originally subtitled "Of Taste" and later "Of False Taste" testifies to the importance of the poem's critique of the concept.

There is one more crucial factor in in the destructiveness of "taste": in order to re-landscape one's estate to follow the demands of taste, one must not only lack "sense" but also have—and spend—a great deal of money. Pope highlights taste's myopic preoccupation with the present in his description of Sir Visto and links it directly to Visto's somewhat suspect fortune: "What brought Sir Visto's *ill got wealth* to waste? / Some Daemon whisper'd, 'Visto! Have a *Taste*.'"[70] Pope draws out the pun on "taste" as a fleeting sensory experience as well as its connection to appetite in order to emphasize the way that preoccupation with fulfilling standards of "taste" shifts attention from the future needs of the land and its inhabitants and toward the momentary demands of sensual gratification. Taste is *only* concerned with the present, and focusing on satisfying the urges of the present, as Pope goes on to show, all too often means severing ties to the past in a way that sacrifices the security of the future. The rhyme on "waste," furthermore, ties taste explicitly to the legal definition of waste: the failure to steward natural resources so that there are "such sufficient Stores as he found when he entered" into ownership,[71] a definition that, as the first chapter demonstrated, had strong implicit and explicit connections to usufruct. Pope also emphasizes the waste of Visto's "ill got wealth," implicitly tying together money, immorality, and the sort of failures of usufructuary stewardship that he associates with "taste." Pope offers no more details on what sort of fortune Visto has, or how he got it, but given the sharp contrast the poem draws between those who waste their money and those who properly steward their land, as well as on the ephemerality of "taste," there is a strong implication that Visto's wealth is money gotten through novel new means. As in the case of Sabinus's son, the preoccupation with the present that drives Visto's concern for "taste" causes him to violate his ethical obligation to conserve and pass along his riches to others and to future generations, and the fact that he has the power to do so because of his new acquisition of money suggests that this sort of new "ill got wealth" is breaking the socio-environmental chains.

In *Burlington*, fashionable aesthetics that frame the landscape as orna-ment rather than a living collective of beings is morally as well as literally barren. This theme is most fully developed in the case of the *Burling-ton*'s arch improvident, Timon. Timon's villa famously unites every flaw of taste in one terrifically uncomfortable estate, where "the suff'ring eye inverted Nature sees," and guests "in plenty starv[e], tantaliz'd in state."[72] Timon's villa offers no comfort, sustenance, or permanence. In William A. Gibson's words, Pope's description of Timon's villa underscores the "tran-sience of such vain displays as his."[73] It is an isolated and isolating place, and not solely because its poorly planned landscape architecture makes its master a "puny insect, shiv'ring in the breeze" that sweeps untram-meled across his too-large downs.[74] As with Visto and Sabinus's son, the physical discomfort and isolation of Timon's estate signifies a more fundamental problem of isolation that underlies his architectural and, particularly, his landscaping choices. The unpleasantness and vast empti-ness of Timon's grounds reflect his failure to "follow Nature" in designing them, not only in the landscape architectural sense but also in the sense that to "follow Nature" is to keep in sight the "natural" interconnection and accountability among beings. Timon's inability to think beyond the aesthetic effect he tries to achieve deadens the landscape by objectifying it: "The suff'ring eye inverted Nature sees, / Trees cut to Statues, Stat-ues thick as trees."[75] Pope disapproved of topiary on aesthetic grounds, but these lines underscore that part of the problem with topiary is the way it reduces trees to inert material, denying them their vital embodi-ment of species and generational interconnection, as well as their abil-ity to offer to the rest of the community the gifts God designed them to give (e.g., shade). Trees for Timon have value *only* for their aesthetic qualities; they are as lifeless in his eyes as stone. Linguistic references to or puns on images and accoutrements of death litter the lines that follow. Timon's gardens boast statues of gladiators as they "fight, or die, in flow'rs," and "swallows roost in Nilus' dusty Urn."[76] The latter func-tions as a reference to the ironic dryness of Timon's water fountains, but also as a pun on the funereal "Urn" used to hold cremated remains. The repeated inversions of life and death in Timon's deadened gardens act as signals of the barrenness of his aesthetics, the moral corruption those aesthetics signal, and, most importantly, the specific way in which his aesthetics have rendered him insensitive to the ways his life, gifts, and duties as a landowner are bound up in the lives, gifts, and duties of the human and nonhuman creatures that his "taste" is incapable of seeing as anything other than objects.

Furthermore, Timon's deadened, disconnected, inverted landscape is made possible by money: "At Timon's Villa let us pass a day, / Where all cry out, 'What sums are thrown away!'"[77] These lines operate on a couple of important levels. First, "all" could refer either to the people imaginatively visiting the villa, or to the actual landscape itself, announcing through its very appearance that the "sums" of money Timon has spent have been wastefully "thrown away." The subtle personification at work in that ambiguous "all" underscores the living, agential nature Pope locates in the estate, as well as Timon's own blindness to it. Second, the place itself naming the changes made to its appearance as wasteful once again connects money, taste, and objectification of landscape. Money enables Timon to put into practice the pursuit of fashionable "taste" that deanimates his estate and severs his own connection to it. "Improvements" such as Timon's are not investments in a place; rather, by valuing his villa for what money can make it into, Timon degrades its true value, and the true value of his own role as a landowner.

The "selfish egotism" that drives Timon to throw away his money on vain, fashionable renovations, James Noggle argues, "fits the profile of the 'moneyed interest'" as J. G. A. Pocock has defined it against the "civic humanism" of figures like Pope's Burlington or Sabinus.[78] For Pocock, the dispute between the "patriot" and the "man of commerce" came down to a question of the relationship between property ownership and political personhood. Property conferred autonomy, and "autonomy was necessary . . . to develop virtue or goodness as an actor within the political, social, and natural realm or order."[79] Unlike money, property did not exist "to engage in trade, exchange, or profit; indeed, those activities were hardly compatible" with virtuous citizenship, because they broke the relationship between lord and land that guaranteed virtue.[80] My reading of Timon through the lens of the usufructuary ethos adds an additional wrinkle to that insight: the sort of virtue Pope defends, and the sort of moral vision he propounds, resist the transformation of property into transferrable goods out of a recognition that an inherent piece of the morality of possession is couched in the interspecies, intergenerational socio-environmental relationships that the estate both fosters and protects. This notion of socio-environmental virtue coexisted with the classical notion of civic virtue Pocock locates at the center of the "landed interest" that emerged in the late seventeenth century. Another reason for the appeal of that notion, I argue, was the neat way it fit together with the coeval usufructuary values this book traces. An important part of the moral threat posed by the "moneyed interest," therefore, lay in the ways it created and exploited fractures in the relationships among human beings and the *living* land they are usufructuaries of. Timon's estate—from its dependence on the "sums" that must be "thrown away" to make it, to its inversion of living nature into inanimate objects, and finally to its projected dissolution—embodies that moral threat and its socio-environmental consequences.

Yet Timon's profoundly immoral misunderstanding of his relationship to his estate has the deepest consequences for himself. His erasure of the living beings and mutual obligations that make up his estate ultimately result, ironically, in his own erasure from history—and the landscape—after his death. The positive side of prodigality such as Timon's was (as was commonly believed in Pope's time) that it inadvertently contributed to the (human) community. Through Timon's grandiosity and false magnificence, his money flows into the human community, and as a result the "Poor are cloth'd, the Hungry fed / Health to himself, and to his Infants bread / The Lab'rer bears: What [Timon's] hard Heart denies, / His charitable Vanity supplies."[81] Though his motivations are flawed and his attention and efforts are misplaced, Timon at least directs support to his human dependents in the form of creating work for them. Nature will not be permanently damaged by him either; it can, and will, reassert itself and regain balance. It does so, notably, in the same way that proper balance was regained in Finch's "Upon My Lord Winchilsea" in the previous chapter: by erasing Timon's alterations from the landscape. In the case of *Burlington*, however, it is nature itself that achieves this return to its ideal state, without the direct intervention of a human hand. Timon's gardens are not long for this world: "Another age shall see the golden Ear / Imbrown the Slope, and nod on the Parterre, / Deep Harvests bury all his pride has plann'd, / And laughing Ceres re-assume the land."[82] The nature Timon had "inverted" here reassumes its vitality, its power, and its temporal connectedness. Rather than a barren, objectified, deadened landscape, Pope personifies the plants he describes using active verbs and divine figures: they "imbrown," "nod," "bury"; through the living and life-giving figure of Ceres, they even "laugh." Pope's omission of human figures from this passage emphasizes both the vitality and the power of nature. This grain is no inert material on which humans exert their will and power, but living things that reflect a vital natural world with a role of its own that it joyfully fulfills.

Finally, the fact that the grain that has reassumed the land will sustain the land's dependents points back to Pope's quintessentially usufructuary understanding of the interconnections among generations and the land. The land itself here once again embodies the lived continuity from past to present to future, a continuity that had been interrupted, but that will be restored once Timon is gone. In the context of the role of posterity in *Burlington* and in the usufructuary ethos, the transience of Timon's villa is important because it draws attention to how profoundly shortsighted it is to dedicate oneself to the demands of "taste." Taste and money disconnect Timon from the future, either England's or his own. Ignore your role in stewarding the landscape, Pope seems to warn, and you risk being effaced by nature when she "re-assume[s] the land" you misappropriated.

Epistle to Bathurst: Miserliness and the Fracturing of the Usufructuary Order

Burlington finds Pope in his most optimistic 1730s mode. The permanent consequences of "Taste" are confined to the Timons of the world, who erase themselves from their own histories. Their prodigality feeds the neighborhood in the meantime, and their poor choices will be subsumed by nature in subsequent generations. The arc of the Popean universe is long, but it bends toward balance, to paraphrase Earl Wasserman.[83] In this, Pope evinces an even deeper faith in the ontological natural hierarchy than Evelyn or Finch had; unlike Finch's, Pope's unfortunately altered landscape returns itself to its original, proper state, and the limited benefits of Timon's prodigality were better than the discord sown by Heneage Finch's mount. Yet Pope's preoccupation with the potential effects of widespread, endemic misuse of riches betrays an ongoing anxiety about the fact that landlords had the power, ability, and, increasingly, the incentive to do whatever they chose with their property, including ignoring their moral duties. And in Pope's eyes, at least, the influx of wealth via new financial channels and the corruption of Walpole's government made such abuse of riches increasingly likely.

While *Burlington* focused primarily on the personal repercussions of wastefulness for the prodigal, *Bathurst* takes a longer view of the cyclical way bad landlords produce bad landlords in order to depict the broader implications of the two main categories of misuse of riches, avarice and prodigality. *Bathurst* is the product of *Essay on Man*–era Pope, and thus Pope avers that ultimately, "Extremes in Nature equal good produce, / Extremes in Man concur to gen'ral use."[84] But though the universe compensates for extremes over the long term, in the short term landlords' deviation from their proper roles cause a great deal of grief. Pope demonstrates this issue by juxtaposing portraits of the very different—but equally destructive—outcomes of a miserly landlord and a prodigal one. The former is a "backward steward for the Poor," which, as Wasserman has shown, was a damning criticism indeed;[85] but the latter, though he may begin by "spouting . . . lavish streams to quench a country's thirst," will end by flooding his dependents with so much that "men and dogs shall drink him till they burst."[86]

The story of Cotta the miser and his son the prodigal illustrates the contrasting brands of immediate destruction each bring about, even as their juxtaposition also illustrates the way that the world achieves overall cyclical balance. Old Cotta, the miser, has been identified by Maynard Mack as a "City merchant who has bought himself a country estate only to invert the whole life that gave such estates their meaning," an intriguing assertion for its suggestion that Pope sees the ignorance and hostility of the emerging new moneyed class toward the traditional usufructuary values of land ownership as a source of the corruption and chaos England had (in his view) plunged into.[87] Cotta's portrait is one of decay and deprivation, as his neglect of public hospitality brings hunger, disappointment, and danger:

> Tenants with sighs the smoakless tow'rs survey,
> And turn th'unwilling steeds another way:
> Benighted wanderers, the forest o'er,
> Curse the sav'd candle, and unop'ning door;
> While the gaunt mastiff growling at the gate,
> Affrights the beggar whom he longs to eat.[88]

Instead of rest and refreshment, those whom the landlord's possession of the land is meant to benefit and support are turned away; the beggar, rather than being fed, is threatened by Cotta's dog, whom he has cruelly starved as well. This is by design, not oversight: "who," Pope's speaker asks rhetorically, taking on Cotta's voice, "would take the Poor from Providence?"[89] The question points to its own answer, resounding and immediate in the context of both the Christian tradition of teachings on charity as well as Pope's own theories of the interconnected natural and social hierarchies: Providence gave the rich man his gifts in order that they may be used to support those who need them.[90] In *Burlington* Pope wrote, "'Tis Use alone that sanctifies Expence," and *Bathurst* implies that proper use alone can sanctify *possession* of riches, too. A little later in *Bathurst* Pope pairs the "Sense to value Riches" with the "Virtue to *impart*,"[91] underscoring the connection between virtue, ownership, and the landlord's inherent mediality, his role as a conduit in the socio-environmental community. Even the miser's misuse of money violates the moral instructions legible in nature, as Courtney Weiss Smith has pointed out: "'Misers' who 'heap up' gold or 'hid[e]' it underground do something unnatural and morally problematic to it. They ought to heed the instructions encoded in the material world."[92] Cotta's moral failure is his belief that what he owns is truly *his*, and more specifically his to *keep*, and his failure to recognize and honor the way that his possession of this property places him at the center of an interdependent chain of lives, for and to whom he is accountable.

Cotta's son "mark'd" his father's "oversight, / And then mistook reverse of wrong for right"—a pattern of alternating vices that Wasserman notes was a longstanding tradition in moral and Christian literature.[93] But though this pattern may "concur to gen'ral use" in the long term, Pope lingers on the particular and serious harms to the lord, his posterity, the public, and the landscape itself that the prodigal's wastefulness brings about. Cotta's son gilds his gluttony and self-indulgence in a patina of patriotism, but if we remove those protestations, Pope's language emphasizes its deleterious effects on the landscape. His "oxen *perish*"—not in his "country's cause," as he claims, nor for his subsistence, but to sate "the capacious Squire."[94] Cotta's avarice inspired him to preserve his forests, but under his son, the "woods recede around the naked seat," and the "Sylvans groan."[95] While the first line links the disappearance of Cotta's woods to Cotta's son's shame and self-exposure, the second pivots to remind the reader that the repercussions are not suffered by Cotta's son alone. His greed and self-indulgence bring untimely death and destruction to the land for which he ought to be accountable, and the destruction has long-term, even permanent repercussions for the living beings—including nonhuman beings—of which he is meant to be a steward. When the son finally "sells his Lands" to pay his debts, it may in one sense be a sign of possible regeneration—he might have sold to someone better able to fulfill the proper role of a landowner—but that possibility is left dangling. Rather, Pope emphasizes the disjunction and fracturing of the social and natural contracts that the Cotta family wrought. Cotta's son ends up broken and abandoned by his "thankless Country" just as he broke and abandoned his own country seat.[96] It is essentially the same fate faced by Balaam later, and indeed by every misuser in the poem. Whether their sin be miserliness or prodigality, Pope emphasizes again and again not merely the ignominy and misery their own lives end in but also the suffering, destruction, and natural and social chaos they leave in their wake.

In the past, critics have argued that there is an inherent contradiction in Pope's moral condemnation of avaricious and prodigal landlords and his conviction that the long arc of the moral universe bends toward self-correcting balance. Since the avarice of one landlord will, eventually, be balanced out by the prodigality of his heir, these critics argue that Pope's cosmology essentially implies that individual cases of avarice and prodigality are simply a given, and ultimately morally neutral.[97] Pope is concerned only with the *ends*, so the *means* are irrelevant; as long as equilibrium is eventually restored, the argument goes, all's well that ends well, and thus his condemnation of avaricious or wasteful use is hypocritical. However, this argument ignores the deeper ontological aspects of landlordly duty and the depth and literalness of human beings' partnership with nature, as well as their duty to emulate it in enacting the divinely imposed obligation to *share*. Although Pope believes that God has built a self-correcting system (so to speak), the excesses of avaricious and prodigal landlords still hurt the beings around them. Eventual equilibrium does not remove the fact that a generation starved as a result of the miser's avarice; as Pope writes in his condemnation of *Bathurst's* miserly Cutler, "numbers feel, the want of what he had."[98] Nor does eventual equilibrium commute the harm that will come when the prodigal's wastefulness strips the land to feed his appetites. God's system will eventually regain balance overall, but that does not absolve the individual landlord of moral culpability for the damage he does by failing to live up to his usufructuary responsibilities in the meantime.

While *Burlington* ends with a triumphal vision of nature subsuming and erasing Timon's mistakes and of what Burlington's proper stewardship could offer to the nation, *Bathurst* ends with a vision of waste and alienation. The final vignette in *Bathurst* is the story of Balaam, a fallen landlord and minister, debased, debauched, broken, alienated from his home, his land, his family, and his nation; his brokenness and the broken relationships and social structures he has left in his wake reflect the nation his corruption helped fracture. Not coincidentally, as many scholars have noted, Balaam embodies "a Whig capitalist in full-length portrait," with all the associations with greed, corruption, ill-gotten material gains, and paper credit that such a figure carried for Pope and his allies.[99] The possibility of cyclical redemption raised with the Man of Ross in the middle of the poem is left unfulfilled, without even a hint of where regeneration could begin—except in Balaam's total self-immolation, and the possibility (meaningfully unspoken) of someone better, someone with "Sense," taking over the wealth he had appropriated. That Balaam ends the poem stripped of his position, his wealth, *and* his posterity is through the lens of the usufructuary ethos particularly significant, and a fitting punishment. Balaam's son "drinks, whores, fights, and in a duel dies"; his daughter's fashionable marriage leads only to a "Coronet, and P-x for life"; and after taking a "bribe from France" to cover his wife's gambling debts, Balaam is hanged, and his wealth and property are "forfeit to the Crown."[100] As in the case of Timon, Balaam's misuse of his gifts and his flouting of his social and natural obligations result in his isolation and erasure. Having failed to use his riches to connect and support other creatures across time and the ontological hierarchy, Balaam himself winds up cut off from both, denied a posterity or a legacy. Yet whereas *Burlington* rendered Timon's erasure in terms of nature's power of regeneration, Balaam is utterly cut off from nature and natural cycles of any kind, and the poem ends without gesturing toward where recovery from the Balaams of the world might begin.

Pope's greater pessimism is also reflected in the muted and indirect nature of the Man of Ross's legacy. Though Pope hints that the Man of Ross might have had children, the poem's focus is on his unique virtue, which is based in the particular individual choices he made to uphold his usufructuary duties to public and posterity. He is commemorated by the persistence of the effects of his virtue, proving by fulfilling the "ends of being" in his particular role to have lived. Yet he does not seem to persist in the public memory outside of Pope's poetry. It is left to Pope to tell his story and to admonish his reader to follow his example, but the overall tenor of *Bathurst*—reinforced by its immediate pivot from the Man of Ross's legacy of virtue to Hopkins's showy, morally bankrupt self-commemoration in marble—expresses skepticism about the probability that others will emulate its ideals. Certainly they *ought* to, in the eyes of Pope and those who shared his perspective. But whether they *will* is considerably less assured.

The Conclusion, in Which Nothing Is Concluded

The simplicity of *Bathurst's* narrative is tempting, even despite its pessimism: lords like the Man of Ross who honor their socio-environmental obligations are good, if rare; lords like Cotta or Balaam, who eschew their socio-environmental duties, are bad, and the political and economic developments that brought them to power are the source of the destruction they wreak. But the truth, in this as in all cases, is more complicated. As Christine Garrard and others have shown, Pope's political and economic beliefs were not as neatly and straightforwardly partisan as the influential narratives introduced by Isaac Kramnick and J. G. A. Pocock suggested.[101] Pope did not argue that profit was *always* bad, that the avenues of financial self-advancement were never morally open to anyone, under any circumstances. For that matter, neither did Philips, or Evelyn, or even Oswald Dykes in *The Royal Marriage*. On the contrary. While deeply troubled by the implications of things like the nascent stock and credit markets, they sought to, in Dykes's words, find a form of "honest Gain," to accommodate accumulation of wealth within the ethical framework of the usufructuary ethos.

Burlington's triumphal conclusion gestures toward how this might be done. Pope defines his ideal landlord as one who preserves the socio-environmental connections among public and posterity, as illustrated earlier in this chapter. He then goes on to add another circle of connection that will arise from those: the global network of colonial trade, linked together by the "rising Forests" that grow "not for pride or show,/but future buildings, future Navies," and by the "Harbours," "public Ways," and other "Imperial Works" Pope urges the king to invest in, all of which support England's growing hegemony in transoceanic colonial trade.[102] *Burlington's* commercial optimism recalls the utopian vision that closed Pope's georgic poem *Windsor-Forest* two decades earlier, a poem that also emphasized the interdependence of lords, landscape, and the ethical use of possession and power. At the end of *Windsor-Forest*, Pope imagined "half" of Windsor's trees "rush[ing]" into England's rivers and oceans in pursuit of treasure, while the "Unbounded Thames shall flow for all Mankind,/[and] Whole Nations enter with each swelling Tyde,/And Seas but join the Regions they divide."[103] Both *Burlington* and *Windsor-Forest* imagine a reconciliation between the socio-environmental estates of England and the burgeoning world of colonial commerce that takes the notion of interconnection central to the usufructuary ethos and applies it to the interconnections forged through transoceanic trade and the literal travel of English trees around the world. Trade, in *Windsor-Forest*, is itself a mode of socio-environmental community in its own right, relying on the bodies of trees to bring together new networks of humans and nonhumans. *Burlington* and *Windsor-Forest* both adumbrate an analogy between the two spheres—the socio-environmental estate and the world of colonial commerce—that lends the moral authority of the former to the latter.

In fundamental ways, these colonialist conclusions contradict the usufructuary ethics that operate elsewhere in these poems. That is precisely the point. These two modes could, and did, coexist in the same writers and poems, as the cultural authority of usufructuary values lingered while material, cultural, and economic transformations often at odds with them sped forward. One piece of the work Pope's *Burlington* did was, in Richard Feingold's words, to "answer an implicit question: under what conditions are the works of empire and of power and of politics good works and not the result of aggressive greed, luxury, and corruption?"[104] Pope answered: those works are good that can be made to conform to the ethical standards of the usufructuary ethos. Yet the strain that remains in *Burlington,* and the uncertainty of resolution in *Bathurst,* noted by Feingold as well as many other critics, suggest that for all his efforts to marry usufructuary with colonialist ethics, a fundamental rift remains that Pope cannot entirely subsume through rhetorical power alone. Pope's unwillingness to look directly at that rift—abetted by his poetic and rhetorical skill—offers us one powerful picture of the ways two separate sets of socio-environmental values could be operative at once, but his poetry obscures the ramifications of the rise of mercantilist colonialism for the usufructuary ethos and environmental history. For that, we must turn to the commercial georgics of the midcentury.

4

Monocultures, Georgics, and the Transformation of the Usufructuary Ethos

O VER THE past two decades, georgic poetry has experienced a renaissance in eighteenth-century studies.[1] While many scholars have
focused on the genre's intersections between empire, economics, class,
ideology, and agriculture,[2] the georgic has more recently been taken up
by scholars of eighteenth-century environmental literature as a form that
not only epitomizes eighteenth-century environmental thought but offers much-needed insights to twenty-first century environmentalism.[3]
David Fairer's formulation of the eighteenth-century "eco-georgic" emphasizes human responsibility to and for the nonhuman world, the interdependence of the environments depicted by georgic, and the genre's
"temporal context," which places humans in the midst of the passage of
time and space, rather than outside or above it.[4] Christopher Loar has recently argued that "British georgic verse offers an eighteenth-century al-
ternative to Latour's paradoxical modernity" in which nature and culture
were scrupulously kept apart, instead "assembl[ing] a social world that
includes both human and nonhuman actors."[5] Loar's Latourian reading assigns to the eighteenth-century georgic the work of merging the
natural and political worlds into a quasi society made up of human and
nonhuman actors. Meanwhile, back on the nonecological side of georgic
studies, Michael Genovese claims that georgic's "admittedly conservative vision of holistic, sociable labor represents a powerful critique of the
'progressive' spread of wage labor and accumulation of capital by owners
unmindful of their own dependence on others."[6] The "connectivity" of

the worlds depicted in georgic, Genovese insists, works against economic individualism and commodity fetishism, creating a deep sociality among people of all classes and species.

An eighteenth-century environmental paradigm; a mode of conceptualizing the interweaving of the sociopolitical and the natural; a body of writing that resists the immiscible transformations of English economy and land use through the nostalgic invocation of the interdependency of humans of all classes and their nonhuman environs: all of these could describe with equal accuracy the georgic or the usufructuary ethos. Fairer's "eco-georgic" echoes the core tenets of the usufructuary ethos, especially accountability and mediality, suggesting that core conceptual connections existed between the poetic genre and the mode of environmental thought.[7] Loar's and Genovese's observations about the ways that georgic entwines people and things into one greater system likewise echoes, in many ways, the usufructuary ethos's socio-environmental functions.

We first saw how georgic and the usufructuary ethos inflected each other in chapter 2's discussion of Dryden's 1697 *Georgics* and Philips's 1708 *Cyder*. Both these early English georgics played up the aspects of Virgil's poem that intersected with the usufructuary ethos, even as the interconnections of agriculture and mercantile economics began to reshape the literal and poetic landscapes they depicted. This chapter maps the confluence of the georgic and the usufructuary ethos through two key mid-century English georgic poems—John Dyer's 1757 *The Fleece* and James Grainger's 1764 *The Sugar-Cane*—to demonstrate that the economic and agricultural changes brought about by the nascent Industrial Revolution and transatlantic colonial trade in turn reshaped the usufructuary ethos itself. *The Fleece* and *The Sugar-Cane* are the two most prominent georgics of mercantile monoculture produced by British poets. They bookend the Seven Years' War, which consolidated England's hold over the Caribbean, extended its dominance of global colonial trade, and marked a key point in England's slow, ambivalent, incremental shift toward expanding British identity to spaces outside the isle of Britain. *Fleece* coincided also with a period of intensifying change in domestic land use, as the mechanization of textile production and the increasing demand for English woolen exports led to more and more "improvements" of private estates. Landowners enclosed commons and evicted tenants to free up more land for pasturage, accelerating a transformation of the English estate and the relationships between landlords, laborers, and land that had begun centuries before.

The midcentury mercantile monocultural georgics studied in this chapter reflect a subtle shift taking place over the course of the century in the way many English writers understood their relationship to their environments in response to emerging economic and colonial transformations. They reflect the increasing tendency, foreshadowed in previous chapters in the anxieties of writers like Dykes, Evelyn, Philips, and Pope, to understand the environment as wealth generating, rather than sustaining; as a site for transformation in the service of a key product, rather than of sustenance for humans and nonhumans.[8] Yet the fact that poets continued to turn to the georgic mode suggests that they desired to see themselves and the transformations they depicted as in continuity with the past, as perpetuating the usufructuary ethics of use that still carried deep cultural authority and informed the understanding of human-environment relationships. The "characteristic quality" of these poems, in the words of Richard Feingold, is "the habit of interpreting, by means of ancient forms of fiction, phenomena which were radically new. . . . To choose the georgic mode is to assert that these new energies and engines are not incompatible with the old virtues and institutions celebrated in the georgic mode."[9] Like Pope's *Epistles* before them—though in starker terms—poems like Dyer's *Fleece* show "the strain of [the] effort" to fit the new with the old "in unresolved problems [and] in conflicting attitudes towards" the poems' subjects.[10] In the process of reshaping those modes and ethics to new cultural, economic, and environmental contexts, however, they were altogether transformed.

Feingold was the first in a series of scholars to observe that the midcentury georgics' slide from "the turning of the soil to the turning of profits" fundamentally altered the genre in a way that both reflected the changes mercantile colonialism wrought and, arguably, foundered the genre at its core.[11] Karen O'Brien has traced georgic's decline in popularity to the corrosive confluence of commercial and colonial influences, which brought about the "demise of the reciprocal, fraternal model of inter-colonial relations which obtained for much of the century," and, with the depiction of slave labor, "broke the association between productive labor and civic virtue" that had been crucial to georgic morality.[12] The trouble for poems like *Fleece* or *Sugar-Cane* arose from the dissonance between the classical and usufructuary values from which they derived their literary and cultural authority—values both poets wished to draw on to bolster their own poetic and cultural authority—and the actual economic and agricultural practices they depicted. It is a dissonance they strained to resolve poetically. That "dialectic of conservation and expansion" is typical of the deep ambivalence Suvir Kaul uncovers at the heart of eighteenth-century poetic defenses of commercial expansion, which suggest "continuity as much as contrast—the link between the past and the future, the complication of forward-looking poetic desire by forms of historical nostalgia," not least of which is nostalgia for traditional, usufructuary bonds of landlordship and obligation that the classical authority of the Virgilian georgic had helped bolster.[13] What midcentury colonial georgics chart, then, is a profound cultural transformation that took place not only in spite of the desire for continuity with traditional (and in particular, classical) cultural, poetic, and socio-environmental values but in spite of active, conscious attempts to preserve that continuity.

The specific way Fleece and Sugar-Cane attempt to preserve continuity with older values, this chapter will demonstrate, is by transferring the locus of socio-environmental connection and value from land and trees to trade and commerce. The economic potential of land and trees was always, of course, a part of English environmental discourse, but Dyer and Grainger make the natural world's contributions to the flows of colonial economy not only explicit, but the main source of nature's value and of social cohesion. Whereas some recent critics have read the poems' depictions of trade's interconnectivity optimistically, I argue that shifting the work of social cohesion from the environment to the economy represents a fundamental change both from earlier instantiations of the georgic and from the usufructuary ethos.[14] Dyer and Grainger apply the language and topoi of the usufructuary ethos and the georgic to enclosure, mercantile monoculture, and colonialism in order to create the sense that nothing fundamental has changed in English values or methods; that the pursuit of wealth, which both poems, but particularly Grainger's, acknowledge is the goal of such work, is not only permissible but positive. While neither breaks entirely with the caution toward use of riches seen in previous chapters, both poems vindicate the pursuit of wealth qua wealth by depicting it as being in line with traditional, usufructuary values. The Fleece and The Sugar-Cane offer especially clear examples of how precisely the material and economic changes of the latter half of the eighteenth century eroded the usufructuary ethos: not by attacking or discrediting it directly, but by incrementally, even unintentionally, transforming it by using its forms and rhetoric to soften and elide the profundity of the changes taking place.

Enclosure, the Workhouse, and the Dislocation of the Usufructuary Ethos in Dyer's *The Fleece*

In his 1757 georgic *The Fleece*, John Dyer attempts to graft what Suvir Kaul has called an "ethics of mercantilism" onto the usufructuary ethos in order to demonstrate the continuity between them, thereby lending his mercantilist worldview the imprimatur of long-standing English socio-environmental morals.[15] Rachel Crawford has claimed that *Fleece* is an "exquisite example of [eighteenth-century English] georgic's failure" because Dyer "teeters between his celebration of the new world of commerce . . . and his yearning for a traditional paternalistic structure figured by the image of the king as a faithful shepherd," fully endorsing neither one nor the other.[16] This section will show that Dyer's "teetering" is not the product of an inability to choose so much as an attempt to forgo choosing by trying to make the former a natural continuation of the latter. Dyer's anxiety to forge such a connection stems from the "constant fear that the world of trading and the wealth it generates will be fatal to the nation," a fear Dyer shared, on some level though not in quite the same form, with the writers explored in previous chapters.[17] The need to show that, in fact, things like wealth, global trade, and enclosure are not fundamental challenges to the structures and ethics of English culture is at the heart of *The Fleece*. In making that case, however, Dyer supplants the central connective role the nonhuman environment played in the socio-environmental ethics of land use in earlier texts with avatars of the emerging world of manufacture and trade. Dyer strives to preserve the structures of usufructuary thought, but the all-important swapping of means and goals—trees for trade, subsistence for wealth—undermines the connection between the social and the natural, and forms cracks in the usufructuary ethos that would, eventually, lead to its decline.

The starkest examples of Dyer's attempt to graft mercantilist agricultural practices onto usufructuary values are found in his depictions of enclosure and workhouses. Dyer disentangles "hospitality" and "posterity" as fibers of social cohesion from their former associations with land and trees, and redefines the commons as environmentally and socially deleterious and the workhouse as the site of social interdependence. In doing so, he erases the "environmental" side of the socio-environmental ethics of earlier poems. Yet he persists in borrowing language and structures of thought that were deeply associated with the usufructuary ethos in the poetry of the early eighteenth century. Dyer depicts the "dreary, houseless, common fields" both as environmentally barren, overused spaces, "Worn by the plough," and as socially destructive spaces that pull apart the fabric of connection among people.[18] As his poetic predecessors had before him, Dyer connects the environmental and the social aspects of his argument, and further claims that common land encourages its various human and nonhuman inhabitants to steal subsistence from each other and from future generations:

> ... none the rising grove
> There plants for late posterity, nor hedge
> To shield the flock, nor copse for chearing fire;
> And, in the distant village, ev'ry hearth
> Devours the grassy swerd, the verdant food
> Of injur'd herds and flocks, or what the plough
> Should turn and moulder for the bearded grain;
> Pernicious habit, drawing gradual on
> Increasing beggary and nature's frowns.
> Add too, the idle pilf'rer easier there
> Eludes detection, when and lamb or ewe
> From intermingled flocks he steals; or when,
> With loosen'd tether his horse or cow,
> The milky stalk of the tall green-ear'd corn,
> The year's slow-rip'ning fruit, the anxious hope
> Of his laborious neighbour, he destroys. (2.118–33)

One by one, Dyer invokes the socio-environmental roles that common lands are meant to play, and claims that common lands in fact destroy what they ought to build. He begins with the most usufructuary of poetic images, the "rising grove" that ought ideally to serve "late posterity," bluntly stating that the commons are a place where no one observes their duty to future generations. Nor do they observe their duties to the "public" of the commons, the other people, human and nonhuman, around them. Dyer depicts communities built around common lands as anti-usufructuary, composed of individuals who each see all that is "common" as there for the taking, without regard for others. Thus the commons becomes a place where "devouring," "pilfering," and "stealing" are endemic. Human beings snatch the "grassy swerd" from the mouths of "injur'd flocks and herds" because they have failed to plan and plant for copses that would provide firewood. Carelessness leaves horses and cows free to eat the crops planted by others—a direct violation of Virgil's instructions—snatching food from the mouths of future humans and animals. And thieves steal sheep to turn the wool to their own profit. All of this follows a description of common lands themselves as "exhausted," "dull," and "barren," tying the environmental decay of the land to the social decay among the beings who live on it (2.116–18). By deliberately and systematically invoking each socio-environmental aspect of land associated with the usufructuary ethos in order to demonstrate that common lands are antithetical to them, Dyer both delegitimizes the tradition of the commons *and* reaffirms the legitimacy of the values of the usufructuary ethos. These socio-environmental values are what we judge a system of land use by, Dyer implies; the fact that commons fail that test makes them a bad system.

What Dyer attempts to do next is to explain why enclosure, workhouses, and colonial commerce fulfill those values *better*. He does so by relocating the concepts of social cohesion and hospitality from the natural spaces they had been bound up with to the physical space of the workhouse, and from there, transferring the ways cross-species community and interdependence had characterized the estate in earlier poems to the ways weaving machines and international trade create mechanisms of human interaction and dependence. This basic insight is not new: Crawford describes Dyer's workhouse as a "macabre version of the aristocratic country house," in which the "recalcitrant poor" will be "redeemed by trade, which tricks out the pre-commercial georgic vision in more fashionable dress" meant to justify the exploitation it describes.[19] Fairer's response to the challenge Dyer's co-option of the georgic for seemingly ungeorgic, anti-environmental ends poses to his theory of eco-georgic is to label Dyer's workhouses "pastoral," rather than georgic—a slippage of genre that temporarily pulls the poem out of the realm of georgic's pragmatic world of struggle and cooperation with nature into the idealized, "emblematic" world of pastoral.[20]

Yet the stakes of Dyer's attempts to transform the spaces of commerce into new, but contiguous, versions of the old spaces that carry out the same values and cultural work are more complicated than either Crawford or Fairer capture, I contend, and have deeper long-term environmental implications than even Kaul's nuanced explanation of *Fleece's* ambivalence conveyed. Decoupling the literal environment of the estate from the socio-environmental values it had formerly represented and assigning them to "trade" and "wealth" begins to sever the moral interconnection between human and nonhuman "publics" that characterized the usufructuary ethos. Dyer's attempt to show that the economic changes England was undergoing were not a challenge to conventional morality wind up, unintentionally but not coincidentally, deconstructing the usufructuary ethos in a way it had never been before.

Having delegitimized the traditional socio-environmental space of common lands as a piece of the estate, Dyer transplants the values of hospitality and interdependence to the workhouse, and from there to the global system of commerce writ large. The workhouse, Dyer avers, is a "public good," and a natural one, akin to "fountains sure, / Which, ever-gliding, feed the flow'ry lawn" (3.100, 93–94). They with "boundless hospitality receive / Each nation's outcasts" (2.78–79), providing the indigent poor (who have, not coincidentally, been displaced by enclosure) with shelter and work, and the nation with a steady stream of exportable textiles. The "fountains sure, / . . . ever-gliding" invoke the quintessentially usufructuary idea of pouring forth the gifts of God in imitation of nature itself—recall Richard Allestree's "Earth" that "coveys the springs through her veins," and Pope's Man of Ross, whose streams pour "thro' the plain / Health to the sick, and solace to the swain." Dyer reframes the capitalistic synergy of labor and trade as enacting the same usufructuary mediality that both nature itself and the socio-environmental landlord had previously.

Further, the central engine of the workhouse, the loom, becomes an emblem of social and commercial interdependence described using some of the same idioms that georgic poems usually reserved to describe the natural chain of being. After describing the specific work that the carpenter, smith, turner, and graver each contribute to the construction of the machine, Dyer writes:

> Various professions in the work unite;
> For *each on each depends.* Thus he acquires
> The curious engine, work of subtle skill;
> Howe'er, in vulgar use around the globe
> Frequent observ'd, of high antiquity
> No doubtful mark: th'advent'rous voyager,
> Toss'd over ocean to remotest shores,
> Hears on remotest shores the murm'ring loom. . . .
> (3.119–26, my emphasis)

The passage arrives in the midst of a detailed description of the steps required to transform wool into cloth—a very georgic sort of minute description of a specific form of labor—but Dyer's particular interpretation focuses on the way that the machine itself, and the human laborers as pieces of a larger machine that runs global commerce, embody the interdependent world of being in which "each on each depend" that had, in earlier georgics, belonged to the system of nature itself. His shift outward to the global scale emphasizes the way that local labor creates an invisible bond among people across the world that itself represents a sort of benevolent interconnection, a common argument for the moral good of transoceanic and colonial trade during the period.

Dyer persistently recasts georgic tropes of interdependence and cooperative labor in nature as examples of the way commerce and trade at the global scale fulfill roles previously played by the socio-environmental estate and landlord. In book 3, he uses an extended simile depicting farm laborers as "sedulous ants . . . / O'er high, o'er low, they lift, they draw, they haste / With warm affection to each other's aid" (3.116, 318–19). The grain they harvest is not going toward their own subsistence, however, but rather is being exported for profit, a process that Dyer once again works to depict as enacting georgic interdependence. The ant-like laborers "conclude / The speedy compact; and, well-pleas'd, transfer, / With mutual benefit, superior wealth / To many a kingdom's rent, or tyrant's hoard" (3.345–48). Dyer's conclusion of this scene is at once celebratory and ambivalent: the workers are "pleased" to export their crops for "mutual benefit" and "wealth," but the trade could have good or bad ramifications, depending on whether it winds up providing a "kingdom's rent" or a "tyrant's hoard." Still, the process of trade and, more particularly, mercantilist agriculture itself is a positive for Dyer; all that labor, that cooperation, that economic interdependence, in his telling, creates the same sort of network of beneficial interconnections that the natural space of the estate provided in earlier writers.

The bee, that most georgic of insects, also serves in *Fleece* as an example of the way labor and trade cooperate to produce the public good on a global scale. Dyer prays that workers will approach their labors

> . . . like the useful bee,
> To gather for the hive not sweets alone,
> But wax, and each material; pleas'd to find
> Whate'er may sooth distress, and raise the fall'n,

In life's rough race: O be it as my wish!
'Tis mine to teach th'inactive hand to reap
Kind nature's bounties, o'er the globe diffus'd. (2.498–504)

What begins as a fairly standard exhortation to work not merely for personal gain but for the betterment of others turns outward at the end of the passage to make a claim for the power of trade to "diffuse" that good over the entire "globe." That line has a double movement, implying both that trade inspires otherwise "inactive" people to go out into the world to "reap / Kind nature's bounties" around the world, and that trade makes it possible to spread both goods (the literal "bounties") and good to new places and people.

However, Dyer betrays some ambivalence toward the "sweets" that bees produce, licensing their production only as long as they also make "wax," a material used to build their home. It is not a coincidence that bees' "sweets," their honey, is golden. Dyer's image of the bees, and his conditional approval of their honey, connects to *Fleece*'s systematic but fundamentally ambivalent attempts to define and justify the conditions under which accumulating wealth is an acceptable goal and outcome within his quasi-georgic, quasi-usufructuary world. *Fleece* is, after all, a poem about the agricultural and commercial processes of growing, reaping, spinning, weaving, and selling one of England's most profitable exports, woolen cloth. Ultimately, the motive for all the activities he describes, from enclosure to workhouses to overseas trade, is to turn a monetary profit. Twenty years on from Pope's *Moral Epistles* and *Horatian Imitations*, the specters of luxury and avarice still haunted English imaginations of wealth. Dyer was anxious, as Kaul writes, to outline an "ethical commercial and public practice" that fit existing moral standards. His challenge was to find a way to make the pursuit of accumulation of profit, so long seen as antithetical to the usufructuary ethos, amenable to it. He did not succeed—indeed, his project was doomed from the outset—but the details of his failure are significant.

"Flakes of Gold": Trade and the Transformation of the Usufructuary Ethos

We can trace Dyer's attempt to fit an "ethical commercial and public practice" to the moral standards of the usufructuary ethos through *Fleece*'s sustained poetic transformation of gold from a symbol of barrenness and miserly accumulation to a symbol for the network of beings connected across the world through the woolen trade. In doing so, he makes fully textual Philips's implicit distinction between a "lawless love of Gain" and a lawful one, and he continues the project begun in his depiction of the workhouses of relocating the social work done by the English environment in earlier poems to new spaces and objects associated specifically with trade and wealth. Dyer commences transformation of gold at the very outset of the poem, when he contrasts the English soil that produces sheep's "locks of price" with Libya's "yellow dust of gold," a periphrasis for sand that evokes both monetary wealth (gold) and impermanence (dust) (1.128, 138). Lest his periphrasis be too subtle, Dyer drives home the connections among gold, greed, and barrenness: the "yellow dust of gold" is "no more food to the flock, than to the miser wealth, / Who kneels upon the glittering heap, and starves" (1.138–39). Meanwhile, greed drives French weavers to acquire their wool "Basely from Albion, by th'ensnaring bribe, / The bait of av'rice, which, with felon fraud, / For its own wanton mouth, from thousands steals" (1.143–45).

Dyer's depiction of the miser starving upon his pile of gold and of avarice as a thief stealing food from the mouths of the deserving are typical, bordering on cliché. But his apparently straightforward condemnation of the pursuit of wealth is belied by the periphrasis for wool that came before the periphrasis for sand: "locks of *price*." It is the very first periphrasis for wool Dyer uses in the poem (before this, he simply uses "fleece"), and it appears as a piece of the first passage in which Dyer sets out to explain why England is the ideal place, climatically, environmentally, and socially, to raise wool. "Locks of price" conveys what English wool literally is (locks of hair) and what makes it stand out—its "price," meaning its monetary value. That monetary value is tied to the superior value of England's landscape and climate as surely and inextricably as Dyer tied Libya's deserts to barrenness and miserliness. The core problem with gold and wealth, then, cannot simply be that gold or wealth, or the pursuit of them, are inherently bad. Dyer is attempting to thread a much finer needle than that.

After decrying greed, avarice, and the barrenness of wealth, and tying them to the unfriendly climates of foreign lands and the enervating influence of "golden mines" (1.153), Dyer pivots to an encomium celebrating England's natural beauty and fecundity. This is the first step in a poem-long process of rehabilitating the significance of gold by attaching it to wool, and through wool to nature and to commerce:

> See the sun gleams; the living pastures rise,
> After the nurture of the fallen show'r,
> How beautiful! How blue th'ethereal vault,
> How verdurous the lawns, how clear the brooks!
> Such noble warlike steeds, such herds of kine,
> So sleek, so vast; such spacious flocks of sheep,
> Like *flakes of gold* illumining the green,
> What other paradise adorn but thine,
> Brittania? (2.164–72)

In Dyer's description, England is vibrant with color and life, an ideal landscape producing idealized georgic images. The "noble warlike steeds" and "herds of kine" could have marched directly from the pages of Dryden's translation of *Georgics III* onto Dyer's "verdurous lawns." The passage consciously, deliberately reminds its readers which generic world we are occupying, the better to spark in our minds the proper generic socio-environmental and usufructuary associations. The final turn in his description, however, marks a departure both from prior georgics and from Dyer's own depiction of gold in the prior lines. In contrast to the barrenness of gold just couple dozen lines above, the "flakes of gold" here are alive, both literally in the sense that they are sheep, and figuratively in the sense that they are as active and vibrant as the rest of the landscape, "illumining the green." On a literal level, the simile describes the effect of sunlight reflecting off the white coats of the sheep, causing them to appear to be glowing, but Dyer's artful attribution of agency to the sheep themselves serves to underscore the way the passage works to attribute life to gold. These flakes of gold light up the landscape and make it possible to better see and appreciate the life surrounding them. This is a sort of gold altogether different from the dead and deadening "yellow dust of gold" or "golden mines." *This* sort of gold, this English gold, provides an anchor and an ornament to English natural and social life.

Yet "gold" still carries its monetary significance, since part of what makes Dyer's simile so apt is the fact that the "flakes of gold" grazing in the fields represent future monetary profit—in the form of golden coinage—for lord and nation. That is also what sets the sheep, as livestock, apart from their neighbors the "warlike steeds" and "herds of kine": sheep are the source of the commodity, fleece, which knits together the newly forming society of post-enclosure England, and knits England together with its trading partners around the world into one harmonious, commercial fabric. By simultaneously tying together landscape, life, commodity, and trade through the poetic periphrasis of "flakes of gold," Dyer reinforces the periphrastic significance of "locks of price" and begins to reclaim and redefine English wealth and profit as being vivifying rather than enervating. Dyer does not go so far as to redefine greed as *good*—he is no Mandeville—rather, he attempts to reframe English pursuit of profit as *not greed*. Just as he works to demonstrate that the new practices of enclosure and the workhouse have the capacity to perpetuate the old socio-environmental values and structures despite their apparently total removal from them, Dyer reworks gold into a living embodiment of the potential for public prosperity.

Dyer's rehabilitation of wealth is a consistent theme throughout *The Fleece*, although not an uncomplicated one. Labor brings riches and riches bring trade and civilization, in Dyer's imagined history, but those same riches bring with them the ever-present threat of a fall into luxury, sloth, loss of trade, and societal decline. That narrative, along with the accumulated significance of book 1's "flakes of gold," is at the heart of book 2's retelling of the myth of Jason and the golden fleece. Dyer links the progress of ancient civilization to the development of a woolen trade: "hence arose their gorgeous wealth; / And hence arose the walls of ancient Tyre," Dyer begins, reaching back to Phoenicia (2.212–13). When Jason and the Greeks arrive at "golden Phasis" (guided across the seas by "golden stars") (2.279, 241), they find a society in decay, a direct result of its people's decadence:

> Jason advanc'd: the deep capacious bay,
> The crumbling terrace of the marble port,
> Wond'ring he viewed, and stately palace domes,
> Pavilions proud of luxury: around,
> In ev'ry glitt'ring hall, within, without,
> O'er all the timbrel sounding squares and streets,
> Nothing appeared but luxury, and crouds
> Sunk deep in riot. To the public weal
> Attentive none he found: for he, their chief
> Of shepherds, proud AERTES, by the name
> Sometimes of kind distinguished, 'gan to slight
> The shepherd's trade, and turn to song and dance. . . .
> (2.284–91)

From one angle, Dyer's tale of Phasis's fall presents a fable of failures of leadership perfectly in line with the socio-environmental role the lord occupies in the usufructuary ethos. The people of Phasis have forgotten the moral and practical centrality of labor, turning their attention instead to grand but useless "pavilions" while neglecting the "crumbling terrace" of their seaport. Phasis's king has forgotten his all-important pastoral roots, which provided the basis for both his nation's (common)wealth and his moral and social role as leader. Phasis suffers from precisely the "diseases of intemp'rate wealth" that the shepherds in book 1 warned of, because their king has forgotten that "king and priest: they also shepherds are" (1.646, 673).

But the true stakes of Phasis's failure in *The Fleece* are not its loss of rural, pastoral virtue to urban immorality, or even, really, its people's move away from pastoral labor. Their fall into luxury is the result of their habitual abuse of wealth, not the wealth itself ("Alas, that any ills from wealth should rise!" lamented Colin the shepherd in book 1 [647]), and the greatest evil of their abuse of wealth, it turns out, is the fact that it results in the loss of *trade*. With the people of Phasis enervated by luxury, the "bold heroes grasp'd the golden fleece" and absconded with it easily. What follows this loss drives home the centrality of trade to Dyer's fable:

> Thus Phasis lost his pride: his slighted nymphs
> Along the with'ring dales and pastures mourn'd;
> The trade-ship left his streams; the merchant shunn'd
> His desart borders; with ingenious art,
> Trade, liberty, and affluence, all retir'd,
> And left to want and servitude their seats;
> Vile successors, and gloomy ignorance
> Following, like dreary night, whose sable hand
> Hangs on the purple skirts of flying day. (2.304–12)

The "trade-ship" and the "merchant" leave Phasis with the golden fleece, and with them go "liberty and affluence," leaving dearth, hunger, and oppression in their wake. All that is typical of midcentury mercantile ideology. What is important here for my purposes is the subtle sleight of hand by which Dyer substitutes trade for agricultural labor as the source of socio-environmental cohesion. All mention of labor vanishes from this passage; the final fall of Phasis has nothing to do with whether or how its people choose to work, but rather with their exclusion from international networks of trade. The very landscape itself mourns this loss, and Phasis's "desart borders" punningly connect its desertion (by merchants) with its land's desertification. In *The Fleece*'s socio-ecology, the possibility that the people of Phasis could continue to farm for subsistence is not even an afterthought. The idea that the nonhuman world lives and even thrives on its own terms, so central to Virgil's georgics, is likewise conspicuously absent. Without the golden fleece, Phasis loses trade; without trade, Phasis is destitute, its land barren, empty, fruitless, purposeless. Economic and environmental catastrophe are simultaneous and inseparable.

The golden fleece becomes the talisman of what Dyer envisions as the successful reconciliation of agricultural production, profit, and trade with social and environmental harmony. He traces its progress through time and space, repeatedly tying images of desolation and nakedness to those places that have lost the woolen trade and images of virtue and fecundity to those places blessed with it. In yet another telling periphrasis, Dyer describes the meadows of ancient Tarentine at the height of its woolen trade as "cloath'd with costly care" (2.338), a description of sheep that once again weaves together ideas of wealth with social and environmental "care" in a way that makes the former utterly inextricable from the latter. The addition of the notion of the meadows being "clothed" implies the nakedness, and by extension the vulnerability and the uncouthness, of land that is not devoted to cultivation specifically for the purposes of trade. This line of argument culminates in Dyer's encomium to "INGENIOUS trade, to clothe the naked world," in which he celebrates the human innovation of turning various natural fibers, from wool to cotton, from raw materials into textiles and then, implicitly, since he is praising "ingenious *trade*" and not "ingenuity," into mercantile goods (2.395). Dyer exhorts his English readers never to forget that they must compete for their share of the market: "Quicken your labours, brace your slack'ning nerves, / Ye Britons; nor sleep careless on the lap / Of bounteous nature; she is elsewhere kind" (2.428–30). Nature bestows her gifts of raw materials promiscuously; what sets a people and nation apart—what makes them worthy of the golden fleece—is their ability to transform those goods into material for trade, and by that means bring wealth to themselves and value to the world.

Thus labor, in *The Fleece*, though it may also bring social cohesion and clothe the world, always circles back around to generating wealth. Dyer's depiction of georgic labor and georgic cooperation with nature, and its

borrowing of usufructuary values of interdependence and care for public and posterity, culminate in his depiction of trade, rather than land, as
the locus of those values and the vector through which those values are
transmitted, and wealth as the most important product of labor and trade.
Again and again, in the second half of the poem, wool and labor are tied
to the generation of wealth, either literal monetary wealth or wealth as a
metaphor for any kind of value. Labor is what transforms the otherwise
valueless fruits of nature into objects for trade: "'Tis toil that makes [raw
materials] wealth; that makes the fleece, / (Yet useless, rising in unshapen
heaps) // A royal mantle" (3.38–39, 43). The attribution of global social
cohesion to trade offers further justification for enclosure and the ensu-
ing relocation of laborers from the fields to the workhouses: The workhouse labors of English textile mills bring "Superior treasures speedier
to the state, / Than those of deep Peruvian mines . . . // . . . Our happy
swains / Behold arising, in the fatt'ning flocks, / A double wealth," by which
Dyer means a fiber that can provide both wealth from trade of cloth and
"cloathing to necessity," a move that once again reinforces the poem's argument that trade and wealth are the proper grounds for public prosperity.

But perhaps most revealing is Dyer's final vision for Britain's place in
the world:

>   'Tis her delight
> To fold the world with harmony, and spread,
> Among the habitations of mankind,
> The various wealth of toil, and what her fleece,
> To clothe the naked, and her skillful looms,
> Peculiar give. Ye too rejoice, ye swains;
> Increasing commerce shall reward your cares. (4.664–70)

Wealth as social harmony, wealth as material support for those in need,
and wealth as monetary profit are merged into one concept, the "wealth
of toil" that Britain will "spread" throughout the world through trade.
The reward that the English people will reap is likewise both monetary
and moral; "increasing commerce" means increased global commercial dominance as well as moral dominance. But even as Dyer works to
convince his reader that trade is at the heart of global social harmony,
he always reverts back to the language of monetary value: wealth, commerce, the *golden* fleece. In attempting to fit the older usufructuary socio-
environmental values to his mercantile world, he cannot escape the fact

that in the world he is describing, value is always and only measured capitalistically. And though the rhetoric of the usufructuary ethos and of the georgic may be carried over, those systems of valuation and of inhabiting the world are fundamentally changed in the transposition.

On some sub- or semiconscious level, Dyer seems to have sensed that fundamental alteration and the threat it posed to his attempts to preserve England against the very threats he describes as having sunk other nations. Immediately after the above passage, Dyer issues a warning about the need to limit wealth: "A day will come" when England will clothe the whole world, but only "if not too deep we drink / The cup, which luxury on careless wealth, / Pernicious gift, bestows" (4.671–73). The moral threat of wealth lingers through the whole poem, as Dyer warns that the pitfalls of the pursuit of wealth must always be protected against. Gold, that "attractive metal, pledge of wealth," is an ambivalent force, a "spur of activity, to good or ill / Pow'rful incentive," which he seeks to harness and channel only for the public good (4.48–50). Yet in shifting the heart of public good and social cohesion to wealth, trade, and commerce, Dyer has made it impossible to disentangle "good" from "gain." The rootedness that earlier versions of the usufructuary socio-environmental English world offered, in which human and nonhuman beings were imbricated in a local, interdependent, though hierarchical, community, finally gives way entirely in *The Fleece* to the more abstract interdependence of the global market. The ethical and environmental consequences of that shift, suggested in *The Fleece*, will become fully clear in the context of the colonial commodity plantation. And so we now turn to our final example, *The Sugar-Cane*.

The Landlord and the Planter in Grainger's *Sugar-Cane*

James Grainger's *The Sugar-Cane* labors poetically to equate the mercantile colonial monoculture of St. Christopher's Island (more commonly known as St. Kitts) with traditional usufructuary values, and to authorize these emergent agricultural practices by their association with the moral and poetic authority of the usufructuary ethos and the georgic. But the underlying disparities between the two sets of practices and worldviews continually undermine his efforts. In this, *Sugar-Cane* is not categorically different from *Fleece*, but its ecological and colonial context throws into especially stark relief the ways in which the changing conditions of colonial agriculture were, in turn, changing both the English georgic and the conceptual relationships the English had with their environments. Like Dyer, Grainger attempts to create continuity between the socio-environmental spaces and figures of the English usufructuary landscape and new social and agricultural structures in the face of the extreme contrasts between their situations. In Grainger's case, however, his efforts focus on attempting to create continuity between the usufructuary English landlord / estate and the colonial planter/plantation. By invoking the trope of the landowner as preserver and protector of the socio-environmental space of the estate, Grainger links his ideal planter figure, Montano, to the figure of the landlord as God's usufructuary, and thus to the entire conceptual-ethical framework of the usufructuary ethos that legitimizes the landlord as the socio-environmental linchpin who guarantees the sustainability of the estate. As Grainger describes Montano literally re-creating his lost ancestral estate on his new colonial plantation, he maps the socio-environmental ethics that licensed the privileged position of ownership and power of the landlord of the English estate onto the colonial plantation in order to elide the differences between the spaces and the material conditions of labor, possession, and power that took place in them.

In this context, it is especially significant that Montano's story follows the trajectory from "subsistence-surplus" farmer to sugar planter that Jason Moore has identified as the standard transition that successful sugar colonies underwent during the early modern period.[21] The "subsistence-surplus" phase of colonial agriculture describes a period during which most planters grew a combination of food crops and cash crops, usually tobacco, before they became financially capable of transitioning to sugar monoculture. Moore notes that the final transition to sugar had major environmental consequences. The larger size of sugar plantations "seems to have been a negative environmental factor in itself,"[22] owing to the need for major transformations to the landscape, such as clear-cutting forests and terracing hillsides. Most of that back-breaking labor was done by enslaved laborers. Montano's story, which follows him from subsistence farmer to subsistence-surplus farmer to wealthy sugar planter, thus stands in not only for the colonial version of the "good landlord" but also for the very processes by which Caribbean islands were transformed ecologically by colonial occupation and plantation monoculture. Grainger's attempts to fit Montano's St. Kitts plantation into the paradigm of the usufructuary ethos thus also provides a glimpse of the way the English attempted to fit the changing environmental, social, and economic contexts of their rapidly expanding colonies to existing environmental beliefs and values—and, at the same time, attempted to deny the profundity of the cultural, ethical, and environmental changes they wrought.

Grainger introduces Montano by connecting him to that most usufructuary of tropes: the grove of trees that binds the socio-environmental community. In one of *Sugar-Cane*'s most conservationist-sounding passages, Grainger decries the practice of clearing forests from sugar-growing islands to stop them from "foster[ing] the contagious blast," a reference to the widespread eighteenth-century belief that trees attracted pest insects to crops.[23] The horror Grainger expresses toward these "Foes to the Dryads" gestures obliquely toward the environmental problems with island tree clearing that were gradually coming to be known: the "Foes," Grainger writes, "remorseless fell / ... each tree of spreading root" (1.560–61), invoking trees' and shrubs' integral role in providing the root structures that prevented catastrophic mudslides and rapid soil erosion experienced on other islands.[24] But Grainger's primary reason for urging the preservation of trees on plantations links Montano, his ideal planter, to the figure of the usufructuary English landlord. The trees, Grainger writes, provide succor to the inhabitants of the plantation during the hottest parts of the day. The choice to preserve them knits together the landscape and the people in a way that invokes the socio-environmental landscapes of earlier writers:

> Ask him, whom rude necessity compels
> To dare the noontide fervor, in this clime,
> Ah, most intensely hot; how much he longs
> For cooling vast impenetrable shade?
> The muse, alas, th' experienc'd muse can tell:
> Oft, oft hath she their ill-judg'd avarice blam'd,
> Who, to the stranger, to their slaves and herds,
> Denied this best of joys, the breezy shade.
> And are there none, whom generous pity warms,
> Friends to the woodland reign; whom shades delight?
> Who, round their green domains, plant hedge-row trees;
> And with cool cedars, screen the public way?
> Yes, good Montano; friend of man was he....
> (1.566–80)

Grainger picks up the familiar pattern here in sketching the connection among landscape, human beings, and the ideal actions of the man who controls the land. "Good Montano," being the good landlord he is, recognizes that preserving living trees supports the lives of the other living beings that share that space ("slaves and herds"), forging a harmonious "woodland reign" whose "hedge-row trees" screen the "public ways," providing relief to the human and nonhuman denizens of the plantation. The echo of the groves described by authors like Evelyn, Finch, and Pope in previous chapters is deliberate: Grainger is painstakingly mapping the socio-environmental functions of the English estate onto his idealized Caribbean plantation. Montano's choices reflect the persistence of the usufructuary ideal that places the landlord at the heart of social and environmental stability. His privileging of trees and their communal rather than monetary benefits frames him as the continuation of the landlord figure who honors his role as the steward of the public and posterity.

The implicit ties to the socio-environmental English estate are only the beginning, however. Grainger creates direct parallels between Montano's lost English estate and his plantation as well, reinforcing the social and ethical continuity of the usufructuary landlord despite geographical contrasts. On his "native shore," from which "persecution" had driven him, "many a swain, obedient to [Montano's] rule, / Him their lov'd master, their protector own'd" (1.581–82, 584–85). Grainger's language not only emphasizes his tenants' affection for Montano but also hints at a sense of interdependence and reciprocity in the relationship—the "obedient" swains "own'd" him, their "lov'd master"—that twists Montano's position into a sort of quasi-usufructuary chiasmus: Montano is at once master and "own'd," echoing the medial position of the usufructuary lord as both authority and subordinate. Describing Montano's relationship to his former tenants in terms of mastery and ownership also sets up and strengthens the parallels to his relationship to the enslaved laborers on his plantation later on, whom he literally owns, but who share the English swains' affection and affiliation: "Well-fed, well-cloath'd, all emulous to gain / Their master's smile, who treated them like men" (1.611–12). Having used "master" in the earlier context softens the contrast between Montano as slave owner and Montano as landlord, as does Grainger's insistence that he treats his enslaved laborers "like men" in the same way he did the swains who had been "obedient to his rule." His generosity and his fulfillment of his duty to use his possessions to support other living beings remain the same: "His gate" on St. Kitts "stood wide to all; but chief the poor, / The unfriended stranger, and the sickly, shar'd / His munificence: No surly dog, / Nor surlier Ethiop, their approach debarr'd" (1.619–22). Montano's story closes with a message to his posterity that emphasizes the moral obligations that come with possession: "Your means are ample," he tells his eldest son, "Heaven a heart bestow!" (1.638). The Montano episode of *Sugar-Cane* is thus bookended by moral admonishments that frame the colonial planter as having the same socio-environmental position and duties as the traditional landlord. This indicates the persistence of the authority of the usufructuary ethos across colonial spaces, as Grainger invokes these values to create an ideal planter for his reader to aspire to.

Yet the troubling conflation of English tenant and enslaved Caribbean laborer in the metaphorical play on "ownership" points toward the deep disparities Grainger is trying to deny. All of Grainger's efforts to portray colonial plantation ownership as the continuation of usufructuary socio-environmental values cannot remove the material and contextual differences between an English estate and a colonial sugar plantation, and those differences challenge the very basis of the values Grainger appeals to. The contrast becomes clear when we compare Montano to his closest georgic exemplar, the Old Man of Tarentum. Virgil's original Old Man of Tarentum had a "few acres of unclaimed land, not rich enough for ploughing, nor fit for pasturage," but by dint of hard work, he managed to "load his table with an unbought banquet" of vegetables and fruits, and "be enriched with teeming bees" that he attracted to his gardens.[25] Virgil plays with "banquet" and "enrich" to point to the true value behind the Old Man's modest subsistence, in contrast to the empty luxury of the city. Likewise, in *Cyder*, Philips set up his own "frugal Man . . . Rich in one barren Acre" which provided him with the subsistence he desired and not more, then immediately contrasted that figure with those greedy landlords who see their land as a way to fraudulently accumulate wealth.

Montano, for his part, begins his time on St. Kitts like Virgil's and Philips's exemplars, toiling on a small plot for subsistence: "At first a garden all his wants supplied, / (For Temperance sat cheerful at his board), / With yams, cassada, and the food of strength" (1.594–96). But within the course of that very sentence, Montano shifts from farming food, to farming cash crops, to overseeing a plantation worked by enslaved people. It is the same progression that Jason Moore tagged as characteristic of Caribbean sugar islands. Grainger frames the progression as the fitting reward for Montano's virtue. In a "neighboring dell" (1.597), Montano planted

> Raleigh's pungent plant, [which]
> Gave wealth, and gold bought better land and slaves.
> Heaven bless'd his labour: now the cotton-shrub,
> Grac'd with broad yellow flowers, unhurt by worms,
> O'er many an acre shed its whitest down:
> The power of rain, in genial moisture bath'd
> His cacao-walk, which teem'd with marrowy pods;
> His coffee bath'd, that glow'd with berries . . .
> Oft, while drought kill'd his impious neighbour's grove.
> (1.599–606, 609)

The contrast between Montano's journey from an exile eking out a subsistence living to the most prosperous (and virtuous) planter on St. Kitts and Virgil's Old Man of Tarentum could not be more stark. As James Gilmore puts it, "while Virgil's old man is happy enough through the satisfaction of extremely modest wants, Montano is happy in the enjoyment of great wealth."[26] Grainger's slide from one into the other, like his persistent invoking of the tropes of the usufructuary ethos throughout the Montano passage, is meant to elide the stark differences between the two situations: Montano's enslaved workers, as discussed above, are framed as being essentially the same as his former tenants; his economic success as a planter, because it is the result of "his toil" and his virtue, is essentially the same as the Old Man of Tarentum's "unbought banquet" and treasure of honey. We might even read Grainger's description of the cotton fields as "many an acre" of "whitest down" as an invocation of Dyer's "flakes of gold," those English fields covered in the white, fleecy, lucrative down of sheep, tying the colonial cash crop to the English one.

Furthermore, like Dyer, Grainger draws distinctions between "good" and "bad" modes of accumulating wealth. The "honest purposes of gain" of good planters like Montano are rewarded with "chearful credit . . . / And well-earn'd opulence" (3.329, 1.491, my italics). The motive of wealth accumulation is not itself bad, and pursued with the sort of balance and compassion Montano shows, its proper rewards are money and the things that can be bought with it. In contrast, Avaro, the antitype of Montano, demonstrates what constitutes the opposite of "honest gain." He starts with

a vast tract of land, on which the cane
Delighted grew, nor ask'd the toil of art.

. . . .

                  But, not content
With this pre-eminence of honest gain,
He baser sugars started in his casks

. . . .

One year the fraud succeeded; wealth immense
Flowed in upon him, and he blest his wiles:
The next, the brokers spurn'd the adulterate mass.
(3.464–75)

Avaro, whose land demands no labor or sacrifice from him, succumbs to the "lust of gain" that Grainger warns against earlier in the poem (1.203) and commits one of the sins of the avaricious agriculturalist that Philips warned against in *Cyder*: mixing his product with an "adulterate mass" of inferior product in order to boost his profit margin. Significantly, in Grainger's formulation, Avaro is punished by the market for his transgression against it. His product found to be compromised and subpar, he loses buyers, and therefore money. It is a moral lesson from Grainger to his planter readers about "good" and "bad" ways to be a commodity-crop grower.

More importantly, like Dyer before him, Grainger takes for granted not only the profit motive of commodity plantations, but also the idea that if profit is pursued in the right way—a way that restrains itself from giving in to greed and "lust of gain"—it is fundamentally good, functioning socially and ecologically in a way in line with more traditional forms of agriculture. The contrast between Montano's "well-earned opulence" and Avaro's ill-gotten gains recalls Dyer's distinction between the nation-building "fleecy wealth" and the nation-destroying "intemp'rate wealth."[27] The connection to Dyer's mercantile georgic is reinforced by Grainger's later reference to "Commerce" as a "chain / To bind in sweet society mankind" (4.351–52), which echoes Dyer's *Fleece* in envisioning transoceanic colonial trade as a force that can "bind" people across space and connect places to one another through the movement of commodities, linked to local landscapes through the georgic, from one spot on the globe to another. Whereas the particular landscape and the living beings of the estate had once provided the socio-environmental basis for community,

Sugar-Cane follows *Fleece* in shifting that role to the abstraction "Commerce." Even with groves of trees present on Montano's estate, providing shade for his human and nonhuman chattel, it is still trade that acts at the greatest force of unity in the poem.

Ultimately, what Grainger attempts to gloss over with the equivalences that he forges in the Montano passage is that Montano's status as an ideal planter derives less from how well he treats his enslaved workers than from his successful accumulation of wealth in the colonial economy. Grainger's single-sentence rush through Montano's years of subsistence farming to the point where the land (mysteriously, passively) "Gave wealth" that enabled him to expand his operation into a true commodity plantation reveals that new center of moral gravity that had begun to emerge in *Fleece*, pulling away from the older georgic and usufructuary ethos, as does Grainger's final flourish of rewarding Montano with "increase / Beyond the wish of avarice." Economic success is the central value at the heart of *Sugar-Cane*, and a central measure of virtue and moral worth, displacing the original georgic moral value of labor itself, which would (as in case of Virgil's Old Man) produce its own reward. Yet, still, the fact that Grainger labors poetically to yoke the new values to the old, the old space of the estate to the new space of the plantation, the landlord to the planter, Montano to the Old Man of Tarentum, evinces a reluctance to let go of the older usufructuary ethos, even as its relevance and authority declines, and an anxious desire to find a way to reconcile it to the new economic and environmental context of the colonial commodity plantation.

SUGAR-CANE'S AVARICIOUS ANIMALS AND THE HOLLOWING OF THE GEORGIC ETHOS

The core problem with Grainger's attempts to rationalize continuities between usufructuary/georgic socio-environmental spaces and the colonial monocultural plantation is that such rationalization enables him (and Dyer, and to a lesser extent Philips) to ignore the most fundamental, most important difference between their georgics and Virgil's. Unlike Virgil's *Georgics*, in *The Sugar-Cane*, humans are not doing what they must do to survive. They are doing what they must do to make a profit. Though the distinction is one that Grainger does his best to elide, it fundamentally alters the relationship between the human planters and the environment they exploit. In the original *Georgics*, as chapter 2 demonstrated, the human and nonhuman inhabitants of the environment were joined by the common pursuit of subsistence that, at times, put them at odds with one another but nevertheless underscored the existential kinship of all beings. In the world of *The Sugar-Cane*, the issues that put humans at odds with nonhumans are categorically different. Nonhumans threaten the crop that represents monetary wealth for the planter, reframing the conflict between humans and nonhumans as an issue of property destruction rather than a shared (if conflicting) struggle to survive.[28] Grainger incorporates the georgic topos of human-nonhuman conflict more explicitly and extensively than either Philips or Dyer did, adapting not only the original *Georgics'* repeated warnings about the threats animals pose to the survival of crops but its framing of that conflict in violent, occasionally militaristic terms. Even more so than in early-century English renditions of georgics by Dryden and Philips, however, Grainger's adaptation of that topos is continually undercut by reminders that the relationships among humans, nonhumans, and the environments they inhabit have been fundamentally altered by mercantile agricultural activity, on both literal and conceptual levels. In attempting to describe the conflicts between planters and animals on St. Kitts in georgic terms, Grainger lays bare the profound socio-environmental changes that the material conditions of colonial agriculture produced, and which his recruitment of usufructuary and georgic topoi had been intended to soften and elide.

To drive home how fundamentally different Grainger's depictions of human-nonhuman conflict are, we will return first to an example from the original *Georgics'* depiction of human and nonhuman beings as simultaneously connected to and in conflict with one another owing to their shared quest for subsistence. The close parallel Dryden draws in book 1 between the mouse's "Garner" and the ant's "wintry Store" on the one hand, and humans' own stockpiles of food on the other, offers an especially concentrated example of the ways humans and nonhumans are tied together through their common drive to try to guarantee future survival, both of the individual and of his or her posterity:

> For sundry Foes the Rural Realm surround:
> The Field-Mouse builds her Garner under ground,
> For gather'd Grain the blind laborious Mole,
> In winding Mazes works her hidden Hole.
> In hollow Caverns Vermine make abode,

> The hissing Serpent, and the swelling Toad:
> The Corn-devouring Weezel here abides,
> And the wise Ant her wintry Store provides.[29]

The "blind laborious Mole" evokes the "hard laborious Kind," human beings, suggesting that moles and humans are alike in working tirelessly for survival, though their work may often be at cross-purposes. More strikingly, Dryden alters Virgil's claim that ants steal grain out of fear of a "destitute old age" to attribute to the insect the power of foresight.[30] The "wise Ant" steals from the threshing house in order to make provision for a coming season of want, just as the wise farmer constructs a strong threshing house to safeguard his winter store from thieves like mice, moles, and ants. Though the animals in this passage are introduced as "sundry Foes" to the "Rural Realm," Margaret Doody notes that we "recognize that they are doing what is in their nature and in their own interest, and momentarily we share their point of view."[31] We share their point of view because Dryden frames their actions and motivations in human terms: the mouse loads her granary exactly as her human counterpart does, albeit with pilfered food. The animals' anxious efforts to ensure survival are analogous to the farmer's. All of these animals compromise human beings' attempts to provide for themselves, but by describing their actions in terms of human habits and needs, Dryden prompts the reader to combine fellow feeling for the pests' hunger and anxiety with the frustration and disgust they might normally provoke. Just like humans and animals, sympathy and antipathy must uncomfortably coexist.

The uncomfortable coexistence enforced by Dryden's personification is further emphasized by the rhymes in the last two couplets of the passage:

> In hollow Caverns Vermine make *abode*,
> The hissing Serpent, and the swelling *Toad*:
> The Corn-devouring Weezel here *abides*,
> And the wise Ant her wintry Store *provides*.

These lines focus on especially unpleasant creatures: vermin, snakes, toads, and in the last two lines, insects. (The "weezel" of the third line refers to a weevil, a small beetle that burrows into granaries and destroys crops.) And yet, despite the relatively unsympathetic creatures involved,

Dryden's rhymes on "abode" and "abide" drive home the parallels between nonhuman and human creatures. An "abode" is a home or a dwelling place, and to "abide" is to dwell or to stay in that home. By repeating the word, Dryden offsets the repellent aspects of these creatures (their "hissing" and "swelling") with the more sympathetic notion that they are, like the farmer, simply building a home. Indeed, rhyming the cozy-sounding "abode" with "toad" mitigates some of the creature's amphibian sliminess. However, in the seventeenth century, the verb "abide" had denotations both friendly and unfriendly. While abide did mean to live or to dwell—especially when used in relation to animals—it also meant to "lie in wait," and even "to await defiantly; to encounter, withstand; or to face, esp. in combat."[32] While the vermin, serpents, toads, and weevils are making a cozy home, they are also making themselves adversaries of human beings. The very act of homemaking is a salvo in the never-ending skirmish between humans and nonhumans. Yet their act of war against humans mirrors the acts of war humans perpetrate against pest organisms throughout the Georgics, often described by Dryden in explicitly military terms. The creatures' attempts at homemaking are destructive to humans' in the same way that human beings' homemaking is inimical to the livelihoods of many species of animals and plants. Dryden's personifications as well as his couplets thus make sympathy and antipathy utterly inextricable from one another.

That fact softens Dryden's earlier framing of the "glutton Geese, and the Strymonian Crane" as "foreign Troops [that] invade the tender Grain," indicating that the conflict between bird and human is a function of the necessity of struggle decreed by Jove at the end of the Golden Age.[33] The need to survive shapes all of these relationships, and in this case it also plays a crucial role in licensing Dryden's use of a military metaphor to describe the birds. The need to survive generates conflict among living things, and requires that some things suffer so that others can thrive. Survival both demands and sanctions inflicting some kind of damage, and the Georgics even depicts "plow[ing] across furrow'd Grounds" to plant grain as "on the Back of Earth inflict[ing] new Wounds."[34] In turn, the fact that growing food is a life-or-death struggle adds to the georgic's poetic decorum. It means that agriculture can be described, without descending into bathos, in terms of war, which in turn lends it a dash of the epic. It may be dirty work, literally and poetically, but the life-or-death struggle to produce a steady supply of food elevates all that digging around in the dirt to something worthy of poetry.

When Grainger replaces the georgic struggle for survival with a struggle for profit, he loses not only that tonal decorum[35] but the fundamental relationship between humans and nonhumans that underpins the original georgics. It is crucial, however, to remember that that loss is wholly unintentional on Grainger's part. He does his level best to maintain the incidents and ideas of the original georgic in his own poem, including both the conflict between human planters and hungry animals who prey on their crops, and the georgic emphasis on careful forward thinking on the part of the planter: "Let prudent foresight still thy coffers guard," he cautions the reader, lest "ravening rats destroy, / Or troops of monkeys thy rich harvest steal" (1.167, 172–73). This is a deliberate and direct echo of the *Georgics'* call for careful protection of the harvest against various natural threats. But Grainger's substitution of "coffers" for granary as that which demands protection stands out starkly. The phantom echo of "granary" (and of the mouse's garner) throws into relief the underlying significance of "coffers": the fact that what is being struggled for in this poem is not future survival of humans and animals, but rather future human wealth. That echo resounds through the passage that follows, altering the significance of the actions of the animals that threaten the cane. Grainger's descriptions are a clear and deliberate echo of the *Georgics'* geese, mice, and Strymonian cranes, yet the fact that the "rich harvest" that Grainger's monkeys "steal" is going to the planter's "coffers" rather than his granary means that the struggle between human and nonhuman creatures in *Sugar-Cane* is imbalanced in a way that the struggle in the *Georgics* is not. Rather than a struggle for control of the foodstuffs that both groups need to survive, the rats and monkeys are after the cane for sustenance, while the planter defends it against them in an effort to increase his wealth.

That alteration in emphasis from crop-as-sustenance to crop-as-wealth also alters the significance of the military metaphor Grainger borrows from Dryden to describe the monkeys. Grainger's "troops of monkeys" invoke the "foreign Troops" that "invade the tender Grain," but the presence of "coffers" in the earlier line, and the fact that the monkeys the "*rich* harvest *steal*," render the battle between planter and apes less of an epic life-or-death brawl and more of a heist carried out by greedy thieves. What in Dryden's *Georgics* had connected humans and nonhumans in a single, noble, necessary pursuit here diminishes the motivations of nonhumans even as it weakens the underlying connections among living creatures the original georgics depict. In *Sugar-Cane*, monkeys and rats plunder the cane for a quick and easy meal (note that neither rats nor monkeys are granted "prudent foresight" as a motivation), and humans must fight back to protect the future fruits of their labors: money. Again, Grainger's intention was surely to follow his georgic exemplars by describing the vexed relationship between the farmer and the creatures who pose a threat to his crop. He is, as Johnson put it, keeping Virgil in his eye, attempting to lend West Indian sugar planting the imprimatur of traditional georgic agrarian and environmental values.[36] Yet the material facts of colonial Caribbean agriculture transform the terms of that relationship, resulting in a fundamentally different set of relationships among humans and nonhuman creatures. It is a state of affairs that Grainger has stumbled into inadvertently as he adapted the georgic to the colonial world, a change from his literary and conceptual models that he could not recognize as such at the time, but which was, perhaps, inevitable, given their different historical and material contexts.

The Sugar-Cane's lapses in georgic decorum and philosophy are a direct result of the place and time it depicts. Grainger turns his poetic focus to wealth in large part because that was the focus of virtually all agricultural activity on St. Kitts. Beyond that, the landscape that the poem describes had been transformed by colonial monoculture, the product of a century or so of rapid cultivation of a nonnative crop, shaped by choices made without a grasp of their long-term ecological outcomes.[37] Among those new environmental factors accidentally let loose by humans are the monkeys and the rats themselves. Grainger returns to a discussion of how to protect the canes from the "monkey-nation" and "whisker'd vermin-race" in the second book, appending two lengthy footnotes explaining the origins of each pest (2.35, 62). "Rats," he explains, "are not natives of America, but came by shipping from Europe," the usual by-product of overseas travel (2.64n.). Monkeys, on the other hand, he blames on the French: "The monkeys which are now so numerous in the mountainous parts of St. Christopher, were brought thither by the French when they possessed half that island. . . . They do a great deal of mischief in St. Kitts, destroying many thousand pounds Stirling's worth of Canes every year" (2.46). Tellingly, Grainger calculates the monkeys' destruction of the cane in economic terms: the loss of thousands of pounds "Stirling," not pounds of cane. It makes sense, in a poem that refers to its title organism periphrastically as "waving wealth" and "thy future riches," that the ecological impact of the monkeys should be expressed according to the bottom line (2.48, 1.435).

These footnotes make it clear that what Grainger's planter is truly battling are the unintended ecological consequences of colonialism and monoculture, the upshot of decisions (such as introducing new species) thought to be minor at the time, but which changed the environmental conditions of St. Kitts in unexpected, and threatening, ways. Grainger's revision of the georgic mode, his revision of the relationships among humans and nonhumans, and the transformations of the Caribbean environment brought about by colonial mercantile agriculture, are thus immiscibly intertwined. The environment and practices Grainger depicts are so fundamentally different that he cannot avoid letting slip the new environmental relationships underlying colonial monoculture. By equating those new relationships with the traditional georgic, Grainger not only gives his poem the imprimatur of classical prestige but is able to lessen the gravity of the changes he depicts. Unsuccessful as his efforts might ultimately be, focusing attention on the seeming similarities between Old and New World made it possible to believe, temporarily, that the differences were superficial rather than categorical, and that Caribbean sugar planting was a continuation of traditional English agrarian life, not a shift away from it. That insistence upon continuity in the face of change is crucial because it occludes perception of changes that, with hindsight, portended generic and environmental failure. It is impossible to avoid a consequence that you cannot or will not foresee.

Conclusion

The Sugar-Cane captures English culture in a pivotal and instructive moment. In the midst of a series of gradual changes that would forever reshape its political, economic, environmental, and moral boundaries, this formally traditional georgic of the relatively new practice of Caribbean island sugar planting attempts to paint a picture of this emerging world in the colors of the old. Indeed, it strains at its seams, in places, to articulate how the usufructuary values that still hold sway apply in a place and to a form of agricultural practice that is irreconcilable with it. In Grainger's and Dyer's attempts to fit the new story of British land ownership and use into the old, we can find important lessons about the ways that shifts in the underlying cultural ways of thinking about fundamental issues like the environment take place. The most important of those lessons is that part of the transition involves adapting and re-adapting new situations to old to allow people to continue to believe that their lives conform to the values their culture espouses, whether or not they do. Over the course of the eighteenth century, the georgic became one major generic mechanism by which that shift was softened and partially elided.

Adam Ferguson wrote that communities "admit of the greatest revolutions where no change is intended."[38] In the case of the usufructuary ethos and the midcentury mercantile georgic, it was the desire to deny the degree of change taking place that led Dyer and Grainger to use its rhetoric to rationalize the changes to socio-environmental relations driven by enclosure, mechanization, colonialism, and enslaved labor. Finally, beyond the implications for the fortunes of the usufructuary ethos in the late eighteenth century, the ways that these poems deployed the still-authoritative environmental rhetoric of the seventeenth and earlier eighteenth century in order to fit emergent capitalist practices of agricultural production into older paradigms is instructive for thinking through the ecologies of twenty-first-century capitalism. What this chapter's insights about midcentury mercantile georgics suggest is that the desire and ability to describe capitalistic modes of environmental relation in culturally accepted pro-environmental terms is a key driver of environmental transformation. Being able to say "nothing is *really* changing" makes it possible to ignore the implications of what you are doing. The same logic arguably drives twenty-first-century phenomena such as "green capitalism," too—though in that case, rather than environmental rhetoric rationalizing change as not-change, environmental rhetoric rationalizes not-changing as change. The problems in Dyer's and Grainger's poems, therefore, are not the problems of the eighteenth century. They are representative of issues that still exist today. They offer a concrete view of, if not human nature, a persistent tendency within Western culture of the past two and a half centuries.

CONCLUSION

The Usufructuary Ethos—Legacies

O F COURSE, the usufructuary ethos did not simply vanish with the rise of the material, social, and literary forces that began to unravel it. It morphed and transformed, and persisted in recognizable shapes through the end of the century, though its influence and its pervasiveness waned. It even crops up, from time to time, in contemporary environmental discourse. By way of conclusion, I will sketch one of its major late eighteenth-century transformations, a couple of places where forms of usufructuary thought has appeared in prominent environmental discourse in the last few years, and consider the promises and pitfalls usufruct offers to twenty-first-century environmentalism.

To sketch the eighteenth-century offshoot of the usufructuary ethos, we must return to where we began: with David Orr's 2016 book *Dangerous Years: Climate Change, the Long Emergency, and the Way Forward,* and its reference to Thomas Jefferson's 1789 letter to Madison about debt, in which he asserted that "the earth belongs in usufruct to the living."[1] It is clear now how drastically an American revolutionary and theist's beliefs would have diverged from those of the usufructuary ethos. What Jefferson perhaps had in mind was a version of the usufructuary idea first articulated by the political theorist Giacinto Dragonetti in *A Treatise on Virtues and Rewards* (1765), which was translated into English by Henry Fuseli in 1769 and was quoted, famously, by Thomas Payne. Dragonetti's declaration that "man is only a traveler on earth, and has no more than the transitory right of usufruct" appears during the course of his argument against the morality of private property held in perpetuity and the accumulation of wealth among a few.[2] Unlike earlier proponents of the usufructuary ethos, neither Dragonetti nor Jefferson is interested in making a theological argument for indebtedness or accountability to God, nor

are they interested in reinforcing the links among past and present beings. Rather, this version of usufruct is more closely aligned with Locke's
Second Treatise, emphasizing the individual rights of the living person,
and framing any action that compromises future people's claims—such
as the accrual of debt—as a violation of their rights.

That notion of the usufructuary right of the living (including those
who will someday be living) to the intact inheritance of a livable environment is central to the claims made by the plaintiffs in the landmark case
Juliana v. United States. The case, brought against the United States by a
group of American minors in 2016, alleges that "climate change violates
their constitutional rights to life, liberty, and property by causing direct
harm and destroying so-called public trust assets such as coastlines."[3]
Legally, there is another link to the eighteenth century: the modern form
of the "public trust doctrine," which dates back to Roman law and which
environmental lawyers argue requires governments to protect privately
held natural resources for the public good, is popularly attributed to
none other than Sir Matthew Hale, who first held that the Crown must
preserve public access to waterways in *De Jure Maris*.[4] The lawsuit hangs
on the assumption that the living have only a "transitory right of usufruct" of the earth, and that therefore they have a duty to preserve it intact for its eventual holders, in this case the children. Controversially, the
children are asking the courts to intervene on their behalf to require
the government to stop anyone who might be doing anything to con-
travene their fundamental right to inherent an environment that they
can live in as comfortably and healthfully as their predecessors have. The
government's failure to protect America's environments, the plaintiffs say,
"willfully prioritize[es] short-term profit, convenience, and the concerns
of current generations over those of future generations"—precisely the
sort of failure Evelyn saw seventeenth-century landlords falling into, and
Pope feared would swallow the English estate.[5]

But there are important caveats to address that separate the sort of
usufructuary vision *Juliana vs. United States* adumbrates and that of the
eighteenth century, and that suggest that for all its promise, usufruct is
no environmental justice panacea. The Jefferson-Dragonetti version of
usufruct that *Juliana v. United States* draws on conceives of environmental sustainability as an issue of injustice of humans against other
humans—and humans only. The environment, in this conception, is a
trove of natural resources, the raw materials that support human life.
Gone is the sense of justice to and community with nonhuman beings

that characterized the usufructuary ethos, imperfect and unequal though its hierarchism might have been. In that sense, this particular legacy of usufruct reflects the technocratic managerialism sometimes imputed to "sustainability."[6] In the case of *Juliana v. United States*, there is a good reason for it: in order to get the government to act decisively on environmental conservation, lawyers must make arguments that courts have the precedent and willingness to recognize, and nonhumans do not have rights in American courts. But the connections between the case and Jeffersonian "usufruct" points to the fact that at least one prominent American strain of usufruct evolved away from the intraspecies and temporally inclusive conceptions of community and accountability that the early modern usufructuary ethos had exemplified.

On the other hand, some Christian leaders have begun to return to the usufructuary tenets offered in the Bible to urge their followers to fulfill their responsibilities as God's creatures and his stewards. In his 2015 encyclical on the environment, Pope Francis cited Leviticus 25:23 in order to refute absolute human ownership of the earth:

> we must forcefully reject the notion that our being created in God's image and given dominion over the earth justifies absolute domination over other creatures. . . . [The Bible] implies a relationship of mutual responsibility between human beings and nature. Each community can take from the bounty of the earth whatever it needs for subsistence, but it also has the duty to protect the earth and to ensure its fruitfulness for coming generations. . . . Thus God rejects every claim to absolute ownership: "The land shall not be sold in perpetuity, for the land is mine; for you are strangers and sojourners with me."[7]

Pope Francis is only the most well known among a host of religious groups and leaders who have articulated similar beliefs and warnings, clustered around the fundamentally usufructuary ideas that the earth does not belong to humans, and that we have a responsibility to care for the nonhuman world for its own sake as well as for the well-being of future creatures, and that we are accountable to the divine and to each other for our failure to do so. Indeed, this version of usufructuary environmental belief never entirely vanished, though it is receiving louder and more urgent attention now. In the nineteenth century, for instance, George Perkins Marsh called for a return to older usufructuary values in *Man and the Natural World*: "Man has too long forgotten that the earth was given to him for usufruct alone, not for consumption, still less for profligate waste."[8] The environmental philosopher and poet Wendell Berry pointed out that Marsh "was invoking [a] biblical tradition" when he stated that humans held the earth in "usufruct alone," one that situates humans and nature in a "moral economy, the standard of which is the health of properties belonging to God."[9]

And yet—the long eighteenth century saw the acceleration of political, social, and ecological developments that would ultimately come to be called the "Anthropocene," some of which the very writers who espoused the usufructuary ethos were themselves directly involved in: large-scale monoculture, colonialism, capitalism, industrialization, the steam engine, the clear-cutting of islands for sugar or cotton plantations, enclosure. The sobering fact is, English writers in the long eighteenth century thought long and hard about precisely the same issues and potential solutions that Orr, Pope Francis, and many other contemporary environmentalists do. As Fabien Locher and Jean-Baptiste Fressoz have pointed out, the eighteenth century's consistent and sincere concern about humans' impact on nature utterly failed to prevent the environmental transformations that have proven so disastrous, a "strange and disturbing fact" that belies the common ecocritical assumption that the Enlightenment's antienvironmental attitudes were responsible for its degradation, and that the right attitude toward the environment could be its salvation.[10] That strange and disturbing fact also belies the suggestion that returning to those ideas now is the source of our salvation. The truth, as always, is far more complicated. Laying out the structure and implications of usufruct as an environmental ethic and its role in English discourse does not merely bring to light a thread of eighteenth-century culture that proves surprisingly conservationist. It also draws attention to parallels with contemporary environmental thought that productively challenge our assumptions about how environmental thought has evolved over the last three hundred years. There is promise in returning to these past ideas, particularly the ways they offered full moral significance to the lives and roles of nonhuman beings in their socio-environmental communities. But only if we return with a difference, and with a willingness resist the urge to look for continuity in the face of massive changes and instead face up to the full weight of our usufructuary responsibilities.

NOTES

1. *OED Online*, s.v. "usufruct."
2. Orr, *Dangerous Years*, 129.
3. Ibid., 131.
4. A few early modern scholars have remarked on usufruct's importance to eighteenth-century environmental thought, though only in passing or in narrower contexts. Clarence Glacken's *Traces on the Rhodian Shore* (1967) and Donald Worster's *Wealth of Nature* (1993) both briefly mention the usufructuary nature of early modern understandings of human dominion over the natural world, but do not delve deeply into the subject. James Tully argues in *A Discourse on Property* that Locke's concept of individual property is essentially a fusion of a usufruct and an English trust, in which ownership is predicated on the fulfillment of the landowner's duties to God, and all land remains a part of the commons (see esp. chaps. 2 and 5). For a full discussion of Locke, usufructuary ethos, and scholarly debates on the environmental potentials in Locke's theory of property, see chapter 2 below.
5. Thomas, *Man and the Natural World*, 21 and 25.
6. Feingold, *Nature and Society*, 2–3.
7. Barrell, *The Idea of Landscape*, 58.
8. Barrell, *English Literature in History*, see esp. 31–40.
9. Fulford, *Landscape, Liberty, and Authority*, 10.
10. C. W. Smith, *Empiricist Devotions*, 25.
11. Ibid., 27 and 24.
12. Most of the works this book discusses—poetic or otherwise—do not differentiate between different classes of nonhuman beings when discussing humans' moral obligations. Thus, while this study has overlaps with and implications for eighteenth-century animal studies, they are not my focus. That said, I owe a general intellectual debt to two particular recent books on animals in eighteenth-century studies: Tobias Menely's *The Animal Claim* and Heather Keenleyside's *Animals and Other People*. Both books explore the figure of the animal and the animal's political voice in the development of late eighteenth-century concepts of political community and personhood.
13. Barrell, *English Literature in History*, 22–23.

14. See Feingold, *Nature and Society;* and Kaul, *Poems of Nation, Anthems of Empire.*

15. Jonsson, *Enlightenment's Frontier,* 15.

16. Fairer, "'Where fuming trees refresh,'" 202.

17. Recently, ecocritics have begun to recognize the ways that ecocriticism's foundational dualism has hampered the field's intellectual growth, particularly with regard to earlier historical periods. See, e.g., Bergthaller, "Introduction," and "'No More Eternal.'"

18. Bate, *The Song of the Earth,* 245, 252.

19. Two of the most influential accounts of the disenchantment thesis besides Bate's books are Merchant, *The Death of Nature;* and McKusick, ed., *Green Writing: Romanticism and Ecology.*

20. Besides Fairer's "'Where fuming trees refresh,'" see Hitt, "Ecocriticism and the Long Eighteenth Century"; Drew and Sitter, "Ecocriticism and Eighteenth-Century English Studies"; Sitter, "Eighteenth-Century Ecological Poetry and Ecotheology," and *The Cambridge Introduction to Eighteenth-Century Poetry;* C. W. Smith, *Empiricist Devotions;* and others discussed throughout this section.

21. Drew and Sitter, "Ecocriticism and Eighteenth-Century English Studies," 229.

22. Karl Kroeber's "green" versus "red" reading in *Ecological Literary Criticism* is probably the most blatant example of this critical practice.

23. Worster's *Nature's Economy* is premised on the contradictory coexistence of "arcadian" and "imperial" attitudes toward the environment in early modern culture. Other examples of the "contradictory" argument include Pickard, "Environmentalism and 'Best Husbandry'"; and Wirth, "'So Many Things for His Profit.'"

24. Hitt, "Ecocriticism and the Long Eighteenth Century," 132.

25. Sitter, "Ecological Prospects and Natural Knowledge."

26. Mikhail, "Enlightenment Anthropocene," 226. One measure of the growing scholarly interest in eighteenth-century environments is the fact that Mikhail's article appeared in a special issue entitled "Humans and the Environment." The issue also includes an earlier version of my argument for usufruct as an eighteenth-century environmental paradigm. See Drew, "'Tis Prudence to prevent th' entire decay.'" A look at conference programs in the field during the years between 2010 and 2019 will show an impressive spike in the number of papers, panels, and conferences dedicated to themes related to nature, environment, climate change, or the Anthropocene. At the time of writing, I know of at least three additional special issues of journals on environmental issues forthcoming, not to mention individual articles and books.

27. Warde, *The Invention of Sustainability*, 158. Both Caradonna's *Sustainability: A History* and Grober's *Sustainability: A Cultural History* also locate the origins of the concept of sustainability in the seventeenth and eighteenth centuries, though neither delves into the early modern history in as much depth or detail as Warde does.

28. Warde, *The Invention of Sustainability*, 6.

29. Bonneuil and Fressoz, *The Shock of the Anthropocene*, 72–73. Grove's *Green Imperialism* thoroughly debunked that narrative, which has nevertheless persisted among environmental scholars outside of the environmental humanities.

30. Bonneuil and Fressoz, *The Shock of the Anthropocene*, 78.

31. Ibid., 36, 171–72.

32. C. W. Smith, *Empiricist Devotions*, 28.

33. Kaul, *Poems of Nation, Anthems of Empire*, 223.

1. The Usufructuary Ethos

1. Howe, *Self-Dedication*, 64–65, original italics. Howe served as domestic chaplain to Oliver and then Richard Cromwell from 1656 to the latter's death. After the Restoration, Howe remained a popular and prominent figure among Nonconformists through his London ministry and publications until the 1690s. See Field, "Howe, John (1630–1705)."

2. As Mark Murphy writes in the *Stanford Encyclopedia of Philosophy* entry on natural law, "Every introductory ethics anthology that includes material on natural law theory includes material by or about Aquinas; every encyclopedia article on natural law thought refers to Aquinas." Aquinas's primary articulation of what constitutes "natural law" appears in part 2, Q. 94 of *Summa Theologica*. "Natural law" seeks to establish a "subset of moral norms" that reflect the order of the universe and "not merely the products or creations of subjective viewpoints" (Kainz, *Natural Law*, xv). "Natural law" is a famously nebulous term, and whether it belongs properly to philosophy or to law often depends on who is writing.

3. Aquinas, *Summa Theologica*, Q. 94, article 6.

4. Aquinas, *Basic Writings*, 780, my emphasis.

5. *The Bible: Authorized King James Version*, 153. For more on eighteenth-century engagements with Lev. 25, particularly Alexander Pope's, see chapter 3.

6. Grotius, *The Rights of War and Peace*, 156.

7. Stillingfleet, *Irenicum*, 31.

8. Ibid., 14.

9. Blackstone, *Commentaries on the Laws of England*, "Introduction." See in particular "Part the Second: Of the Nature of Laws in General," 38–41.

10. Pufendorf, *The Law of Nature and Nations*, 358.

11. Ibid., 362. For a fuller discussion of Pufendorf's views on the morality of human use of animals, see Buckle, *Natural Law and the Theory of Property*, 92 and nn. 147, 148. The phrase "moral community" is Buckle's.

12. Locke, *Two Treatises of Government*, 271.

13. Locke writes that "the goods of fortune are never so much ours that they cease to be God's: that supreme Lord of all things can, without doing any wrong, give of His property to anyone as He pleases." *Essays on the Law of Nature*, 203.

14. Locke, *Two Treatises*, 271. There is more to be said on the subject of Locke's connections to the usufructuary ethos as a whole, which I will return to later in this chapter.

15. Howe, *Self-Dedication*, 68.

16. Field, "Howe, John," 473. Obviously, Howe was not without critics, among them Defoe, who addressed the preface of the 1701 reissue of *Occasional Conformity* to him, challenging Howe's public endorsement of the practice (ibid., 474).

17. Stranks, *Anglican Devotion*, 143–44.

18. By 1790 *Whole Duty* had reached its twenty-eighth edition, owing not only to its pervasive influence on Anglicanism but also to its widespread acceptance and use among Dissenters. For instance, John Wesley recommended the book despite its weakness on the question of justification by faith. See Stranks, *Anglican Devotion*, 125.

19. Allestree, *The Art of Contentment*, 22, my emphasis.

20. Allestree, *The Gentleman's Calling*, 394.

21. Ibid., 413.

22. Ibid., 394.

23. McGee, "Adams, Thomas (1583–1652)." Adams spent most of his career in London as the rector of St Benet Paul's Wharf and St Benet Sherehog. He published numerous sermons in the 1620s and 1630s, which were generally well received and widely respected.

24. Adams, "The Barren Tree," in *Five Sermons*, 27. "The Barren Tree" was preached on October 26, 1623, at St Paul's Cross and dedicated to John Donne, then the dean of St Paul's. The edition from which all subsequent quotes from Adams are taken was printed in 1626, but it was composed of independent editions of each sermon printed by the publisher, John Grismand, at other dates. "The Barren Tree" has its own title page dated 1623.

25. Ibid., 23.

26. Allestree, *The Gentleman's Calling*, 414 and 416.

27. Allestree, "Sunday XIII," in *The Whole Duty of Man*, 105.

28. *OED Online*, s.v. "tenure," def. 2.

29. Hales, *The Works of the Ever Memorable Mr. John Hales*, 156, my emphasis.

30. *OED Online*, s.v. "emphyteusis."

31. *OED Online*, s.v. "farmer, n. 2," defs. 2–4.

32. Cromartie, "Hale, Sir Matthew (1609–1676)." Hale's modern reputation is somewhat more mixed. He is perhaps best known today for sentencing two women to death in 1662 for witchcraft (a trial that went on to provide precedent for the Salem witch trials), as well as for asserting in *Historia Placitorum* that a husband could not legally be said to rape his wife.

   Incidentally, Hale is also frequently credited in contemporary American environmental law with a key role in establishing the so-called public trust doctrine, which asserts that the government holds certain public lands in trust for the benefit of all citizens. The PTD is cited by environmental lawyers to claim that the government has both a right and a duty to preserve natural resources for the benefit of citizens. Hale's role in framing the modern PTD has been contested by legal scholars on the basis of both the legal relevance of English precedent to American law and the question of whether his claims for public rights in *De Jure Maris* can be taken as a statement of public trust at all. The legal argument associated with the PTD in *De Jure Maris* has no relevance to his arguments for human usufructuary lordship elsewhere in his writings. For more on Hale and the PTD, see Sax, "The Public Trust Doctrine in Natural Resource Law"; and Blumm and Wood, *The Public Trust Doctrine*.

   For a full biography of Hale, see Cromartie, *Sir Matthew Hale, 1609–1676*.

33. Hale, *Primitive Origination*, 370.

34. From William Smith, *A Dictionary of Greek and Roman Antiquities:* "VILLICUS (ἐπίτροπος in Greek writers, Plut. Crass. 4), a slave who had the superintendence of the *villa rustica*, and of all the business of the farm, except the cattle, which were under the care of the magister pecoris (Varro, R. R. I.2). The duties of the villicus were to obey his master implicitly, and to govern the other slaves with moderation, never to leave the villa except to go on market, to have no intercourse with soothsayers, to take care of the cattle and the implements of husbandry, and to manage all the operations of the farm (Cato, R. R. 5.142). His duties are described at great length by Columella (XI.1, and I.8), and those of his wife (villica) by the same writer (XII.1), and by Cato (c143)."

35. Hale, *The Great Audit*, 12.

36. Eleven separate editions of *Contemplations Moral and Divine* appear in *Early English Books Online* dated between the book's first edition in 1676 and 1700; *Eighteenth Century Collections Online* has another six dating from 1705 through 1792.

37. Hale, *The Great Audit*, 18–19.
38. Kidder, "The Life of the Reverend Anthony Horneck, D.D.," v–vi. For more on Horneck, see W. R. Ward, "Horneck, Anthony (1641–97)," *Oxford Dictionary of National Biography*, 155–56.
39. Horneck, *The first fruites of reason*, 19–20.
40. *OED Online*, s.v. "audit, n.," defs. 1, 3, and 4.
41. Hale, *The Great Audit*, 13.
42. Ibid., 13–14.
43. *OED Online*, s.v. "arrear," defs. 4.a. and 7.a., and "insuper | in super."
44. The *OED* entry on "insuper" includes this sample definition by legal and political writer Thomas Manley from 1672: "*In super*, Is a word used by Auditors in their Accounts in the Exchequer, when they say so much remains *in super* to such an Accountant, that is, so much remains due upon such an Account."
45. Hale, *The Great Audit*, 14.
46. Ibid., 16–17.
47. Ibid., 15–16, 16.
48. Thomas, *Man and the Natural World*, 155.
49. Tryon, "Preface," *The Country-Man's Companion*.
50. Hale, *The Great Audit*, 15.
51. Allestree, *Gentleman's Calling*, 393.
52. Ibid., 413–14.
53. Worster, *The Wealth of Nature*, 97. A. C. Grayling notes in *The Age of Genius* that Louis XIV took to heart this doctrine as a part of his absolutism, since it technically gave him the right to take back his lords' land and re-rent it back to them whenever and however he wished—though he did not take advantage of that putative power (286). In England, the assumption that possession was usufructuary had very different significance.
54. Holdsworth, *A History of English Law*, 5:22.
55. Cowell, *Institutes of the Laws of England*, 97–98.
56. Blackstone, *Commentaries*, 105.
57. Ibid.
58. Cowell, *Institutes of the Laws of England*, 98.
59. An estate for life was roughly what it sounds like: the possession of land and the rights to profit from it for the term of one person's life. At the lessee's death, the estate for life ceased, and the property reverted to the possession of the lessor. In an English common law use, property was enfeoffed to a trustee or trustees A to the use of person B, at whose death the use would be passed to person C, and so on. Other examples of legal writers comparing English property forms to usufruct include Giles Jacob, *A Treatise of Laws: Or, a General Introduction to the Common, Civil, and Canon Law* (1721), who states that an "Estate for Life, Years, or at Will, &c.

here in England, are almost of the same nature as these Usufructs amongst the Romans" (380). Legal writers who compared the English use to usufruct include Sir Jeffrey Gilbert, Chief Baron of the Exchequer (1725–26), in *The Law of Uses and Trusts* (1734); and Blackstone. See Holdsworth, *A History of English Law*, 4:410 and n. 1.

60. *The Landlord's Law*, 150, my emphasis. *The Landlord's Law* went through eight editions by 1739.

61. Ayliffe, *A New Pandect of Roman Civil Law*, 317n. Ayliffe begins his footnote on usufruct and English common law by noting the uncertainty surrounding the legal status of the Crown's ultimate ownership of all land: "As no one here in *England*, besides the King, had a full Property in real or immoveable Estates, it is not easy to discern who are the Usufructuaries, and who are not, *viz* whether all Persons are such, that hold Lands by any Title whatsoever, or only some that hold them by this or that Right." He goes on to conclude that anyone subject to an action of waste, in any case, is essentially the same as a usufructuary.

62. This is the same argument made by A. J. McLean in "The Common Law Life Estate."

63. On Hardwicke and the doctrine of equitable waste, see Holdsworth, *History of the English Law*, 7:278 and 12:259. Philip Yorke, Earl of Hardwicke, was Lord Chancellor 1737–56.

64. Locke, *Two Treatises*, 271.

65. Ibid., 287–88.

66. Ibid., 288, 290.

67. Ibid., original italics.

68. Quoted in Tully, *A Discourse on Property*, 122.

69. Stein, *The Roman Law in European History*, 108.

70. Domat, *The Civil Law in Its Natural Order*, 192.

71. Judge, "Restoring the Commons," 332.

72. Kristin Shrader-Frechette has demonstrated that since Locke's provisos are part of the natural law from which civil society and individual appropriation arise, they hold for all time, and thus operate as a universal limitation on individual property rights and land use. See "Locke and Limits on Land Ownership." Other important contributions to the argument for a Lockean environmentalist ethic include Squadrito, "Locke's View of Dominion"; Elliot, "Future Generations, Locke's Proviso and Libertarian Justice"; Wolf, "Property Rights, Human Needs, and Environmental Protection"; Brock, "Future Generations, Natural Resources, and Property Rights"; Duncan, "Property as a Public Conversation"; and Judge, "Restoring the Commons."

73. Tully argues that Locke's concept of individual property is not plain usufruct, but rather a hybrid of a usufruct and what Tully calls the "uniquely

English concept of the *use*," which, he argues, "serves to underline the major point that proprietorship exists for, and is conditional on, the performance of positive duties to God" (*A Discourse on Property*, 122). Tully's comparison of Lockean property rights to usufructs and English trusts is a piece of his broader argument (which does not touch on environmental themes) that property remains fundamentally part of the commons in Locke even after an individual appropriates part of it to him- or herself through labor. Tully's thesis has attracted several lengthy and detailed critiques by fellow Locke scholars since it was published in 1980—a fact that reflects its brilliance and erudition as much as its flaws. See Neal Wood, *John Locke and Agrarian Capitalism*; Zuckert, *Natural Rights*; and Waldron, *The Right to Private Property* and *God, Locke and Equality*.

74. Wolf, "Contemporary Property Rights, Lockean Provisos." Wolf's article prompted an exchange with Gillian Brock in the journal *Ethics and the Environment*. See Brock, "Future Generations, Natural Resources"; and Wolf, "Property Rights, Human Needs."

75. Trachtenberg, "John Locke," 107. Trachtenberg uses "usufructory" rather than "usufructuary" throughout his essay.

76. Ibid., 110.

77. For a full account of this shift, see Haakonssen, *Natural Law and Moral Philosophy*.

78. Buckle, *Natural Law and the Theory of Property*, 47.

79. Ibid.

80. Trachtenberg, "John Locke," 111–12.

81. Locke, *Two Treatises*, 300, original italics.

82. Ibid., 300–301, original italics.

83. Ibid., 302.

84. Ibid., 301.

85. Lebovics, "The Uses of America," 574.

86. Ibid., 574–75, 575. See also Jimmy Casas Klausen, "Room Enough." Klausen argues that because "consent and natural liberty rely on the availability of open space"—consent to a system can only logically exist if the person consenting can theoretically choose to exit the system—"Lockean liberalism justifies, maybe requires, settler colonialism."

87. Locke, *Two Treatises*, 301.

88. Lebovics, "The Uses of America," 578.

89. I am indebted throughout this section to Dwight Codr's nuanced complication of this history of "finance" in the early eighteenth century, particularly his theorization of the ethics of "anti-finance" and its resistance to the prioritization of profit and certainty. See Codr, *Raving at Usurers*.

90. Sir Roger L'Estrange (1616–1704) was a Jacobite, Tory pamphleteer, and surveyor of the press during the Restoration. Given that Dykes's

publishing career stretches at least into the 1720s, it is most likely he worked as L'Estrange's amanuensis at the very end of the latter's life. Dykes claims to have been "falsely reported to be a Papist," which most likely would have been the result of Jacobite sympathies, though it could also have been a result of his public association with L'Estrange. That Dykes shared at least some of L'Estrange's political loyalties is clear not only from his employment but also from the dedication to his 1708 *Moral reflexions upon select British proverbs . . . after the Method of Sir Roger L'Estrange's Aesop*, in which he laments the "miserable Hardships I have undergone, through difficult times of Revolution, and the troublesome Turnings of State."

91. Taken from a section of the Old Testament dedicated to describing the good wife, the verse reads: "She considereth a Field, and buyeth it: with the Fruit of her Hands, she planteth a Vineyard." Quoted in Dykes, *The Royal Marriage*, 218.

92. Ibid., 220, 220–21.

93. Chapter 4 will return to examine this issue in depth through the midcentury georgic poems of John Dyer and James Grainger.

94. Dykes, *The Royal Marriage*, 220–21.

95. Ibid.

96. Ibid., 223–24.

97. See esp. Pocock, *Virtue, Commerce, and History*.

98. Wycherley, *Miscellany Poems*, 195, 194. Citations of this poem are by page rather than line number.

99. Ibid., 196.

2. Trees, Posterity, and the Socio-Environmental Landlord

1. Allestree, "Sunday IV," *The Whole Duty of Man*, 55.

2. Allestree, *The Gentleman's Calling*, 397, original italics. *The Gentleman's Calling* was first published in 1660, attributed to the (anonymous) "Author of *The Whole Duty of Man*," and was subsequently reprinted as a part of the *Works* and on its own at least twenty times between 1660 and 1696.

3. Pocock, *Virtue, Commerce, and History*, 48.

4. In David Fairer's seminal essay on the environmentalism of the eighteenth-century georgic, "'Where fuming trees refresh,'" *Cyder* provides a key example of the eco-georgic's emphasis on experimentation, mixing, and careful attention to the needs and responses of nonhuman nature to human intervention. Still, *Cyder* receives fairly little space in Fairer's essay. Furthermore, his aim of defining the eighteenth-century "eco-georgic" as an exemplary environmental genre leads him to brush aside

the "socio-political implications of Philips's [...] poem published just two years after the Act of Union" (208–9).

5. E. P. Thompson, *The Making of the English Working Class* and *Customs in Common*; John Barrell, *The Idea of Landscape*; Raymond Williams, *The Country and the City*; James Turner, *The Politics of Landscape*. The critical literature on the subject of the transformations of early modern landscape and of its representations is extensive and certainly not limited to those scholars. More recent important interventions on this topic include Andrew McRae, *God Speed the Plough*; Sharpe and Zwicker, eds., *Refiguring Revolutions*; and Maclean, Landry, and Ward, eds., *The Country and the City Revisited*.

6. John Philips, *Cyder*, 1.436.

7. C. W. Smith, *Empiricist Devotions*. See in particular chaps. 1 and 3.

8. See chapter 1, under the heading "Middle Man: Usufruct and Hierarchy in Seventeenth-Century Devotional Writing."

9. Thomas Adams, *Five Sermons*, 24–25, 25–26, 26.

10. *The Bible: Authorized King James Version*.

11. Allestree, *The Works of the Learned and Pious Author*, 413, 413–14.

12. Ibid., 416, 415, 417.

13. Ibid., 416.

14. Swift famously destroyed his sermons after he preached them, but a few, including "On Mutual Subjection," were rescued by his friend Thomas Sheridan, so that Sheridan could re-preach them in his own pulpit: "One day in particular [Swift] brought above thirty of his sermons from the study into his bed-chamber, where he was going to throw them into the fire; and being asked, what papers they were? He answered, 'old sermons, which I shall never preach again!' Upon which Dr. *Sheridan* begged them, saying, 'they would be very useful to him, who might preach very often;' on which the Dean gave them to him. Three of these sermons, to wit, on the Trinity, on the Testimony of Conscience, and on Mutual Subjection, were published after Dr. *Sheridan's* death, by his eldest son Thomas" ("The Life of Jonathan Swift, D.D.D.S.P.D.," in *The Beauties of Swift*, xxxii). "On Mutual Subjection" first appeared in *Three Sermons* (London, 1744) and was reprinted numerous times through the century in various miscellanies, editions of the works of Swift, and (excerpted) in *The Beauties of Swift*.

15. Swift, *Four Sermons*, 9.

16. Ibid., 3.

17. Two of the most widely cited sources for that critique are Lynn White Jr., "The Historical Roots of Our Ecologic Crisis," which implicates Christianity as being fundamentally anthropocentric and anti-ecological, and Carolyn Merchant, *The Death of Nature*. Numerous other foundational works of ecocritical theory have been premised on the so-called

disenchantment of nature as a result of the scientific and intellectual transformations of the Enlightenment, including Jonathan Bate, *Song of the Earth*; Mark Lussier, *Romantic Dynamics*; and Michel Serres, *The Natural Contract.*

18. A few critics over the years have noted the ecological potential in the Chain of Being's interdependencies, particularly in the context of Pope's *Essay on Man*; see in particular Sheridan Blau, "Pope's 'Chain of being'"; Wendell Berry, "Poetry and Place"; and David W. Gilcrest, *Greening the Lyre*. John Sitter tied the ecological potential of Pope's Chain of Being to eighteenth-century theological thought in "Eighteenth-Century Ecological Poetry and Ecotheology." I will return to Pope's instantiation of the Chain of Being in greater detail in chapter 3. More broadly, innovative recent work by Courtney Weiss Smith, Tobias Menely, Heather Keenleyside, David Fairer, and others has brought to light previously occluded ways in which eighteenth-century English writers perceived their imbrication in a vibrant, living natural world.

19. Johnson, *Dictionary of the English Language*, s.v. "steward." Quotation appears here as printed in the *Dictionary*, which is slightly altered from Swift's original.

20. Allestree, *The Works of the Learned and Pious Author*, 413, my emphasis.

21. Ibid., 420.

22. Flavel, *Husbandry Spiritualiz'd*, 285. Courtney Weiss Smith has shown that Flavel "formalized the way that" long eighteenth-century devotional literature moved "smoothly back and forth between and among parallel or simultaneous registers" of literal and spiritual. See *Empiricist Devotions*, 45–47.

23. Flavel, *Husbandry Spiritualiz'd*, 290, my emphasis.

24. Rival, "Trees," 19. Rival cites Simon Schama's *Landscape and Memory* when she makes this claim, but Schama's chapter on early modern English forests—which includes a long section on Evelyn—is uncharacteristically wrapped up in trees' political and economic significance, and does not make note of the social and environmental significance of trees I uncover in this chapter.

25. This chapter examines the last version of the book that Evelyn had his hand on, the 1706 *Silva*. I will refer to the book as "*Silva*" throughout.

26. Hartley, "Exploring and Communicating the Knowledge of Trees," 229–30.

27. Darley, *John Evelyn*, 181–82. For a more detailed history of the destruction of England's forests under the Stuart kings, see Schama, *Landscape and Memory*, 154–58.

28. Evelyn, "To the Reader," in *Silva*.

29. Hale, *Primitive Origination*, 370; see chapter 1 above.

30. Chambers, *The Planters of the English Landscape Garden*, 44.

31. Evelyn, "To the Reader."
32. Evelyn, *Silva*, 288 and 310.
33. Ibid., 68.
34. Ibid., 22.
35. Ibid., 9.
36. Abraham Cowley, "The Garden," in *Silva*, n.p.
37. Swift, "On Mutual Subjection," in *Four Sermons*, 9.
38. Grove, *Green Imperialism*, 56–58. Chapter 4 will return to the challenges colonial spaces posed to the usufructuary ethos.
39. Evelyn, *Silva*, 1–2.
40. All references to Heneage henceforth are to Heneage, second Earl of Winchilsea, Charles's grandfather and Anne's father-in-law, *not* to her husband Colonel Heneage Finch.
41. Finch, "Upon My Lord Winchilsea," in *Poems*, 1–6. Further citations to this work are in the text.
42. Hamrick, "Trees in Anne Finch's Jacobite Poems of Retreat," 549.
43. Nicolle Jordan, "'Where Power Is Absolute,'" 262–63.
44. Ibid., 261.
45. Ibid., 258.
46. Ibid., 266.
47. Evelyn, *Silva*, 295.
48. McKusick, "John Evelyn: Forestry of Imagination," 110.
49. Qtd. in ibid., 110.
50. I am indebted to J. C. Pellicer's introduction to the 2001 edition of *Cyder* for pointing to Evelyn as a source for Philips. For Pellicer's references to *Silva*, see "Introduction," iii and viii.
51. As Shaun Irlam puts it in his study of Grainger's *The Sugar-Cane*, "the considerable classical prestige of the georgic form was marshaled during the eighteenth-century to produce, stabilize, and legitimate an agrarian-capitalist organization of English and colonial countrysides." See "'Wish You Were Here,'" 378.
52. Pellicer, "Introduction," xii. Pellicer here explicitly emphasizes the "partisan and specific" politics of the poem over the "broader lines of 'ideology' as commonly understood in socioeconomic terms." Specifically, he argues that *Cyder* was a piece of Robert Harley's late 1707/early 1708 attempt to forge a broad Tory coalition that was, paradoxically, published by the Whig Jacob Tonson. Pat Rogers has made a case for *Cyder* being a Tory response to the Act of Union, and Robert P. Irvine sees the poem more broadly as an attempt to assert the moral and political centrality of land against Lockean political economy. See Pellicer, "Harleian Georgic from Tonson's Press"; Rogers, "John Philips, Pope, and Political Georgic"; and Irvine, "Labor and Commerce."

53. Rogers, "John Philips, Pope, and Political Georgic," 414.

54. Philips, *Cyder*, 2:117–23.

55. Ibid., 2:368–70

56. Ibid., 2:137–39.

57. Ibid., 2:146–53, 161.

58. Prideaux, *The Original and Right of Tithes*, 17. Prideaux summarizes this argument as the most common one leveraged in defense of traditional tithing, then devotes the bulk of his treatise to an elaborate dismantling of it based on his own interpretation of the Old Testament. A Whig and a Low Churchman, Prideaux was dean of Norwich from 1702 until his death in 1724. He was a vocal opponent of Catholicism and of James II, and supporter of William and of adjusting Anglican doctrine and practices to accommodate Dissenters. His argument against tithing in *The Original and Right of Tithes* was an important piece of this legacy, but he is most famous as a writer for his anti-Islamic books, *Life of Mahomet* (1697) and *Connection: The True Nature of Imposture Fully Displayed in the Life of Mahomet* (1716–18). See Quehen, "Prideaux, Humphrey (1648–1724)."

59. Philips, *Cyder*, 1.656, 660–61.

60. Ibid., 1.595.

61. Ibid., 1.643–48.

62. Virgil, *Georgics*, in *Virgil I*, 109.

63. For a more detailed discussion of this dialectic, particularly in Dryden's translation of the *Georgics*, see Drew, "'Iron War' as 'Daily Care.'"

64. Virgil, *Georgics*, 163.

65. Ibid.

66. Ibid., 199, 173.

67. Dryden, *Georgics*, 2.739–40, 743–44, my emphasis.

68. Virgil, *Georgics*, 173.

69. Goodman, *Georgic Modernity and British Romanticism*, 1.

70. Virgil, *Georgics*, 133.

71. Ibid.

72. Dryden, *Georgics*, 4.77–82.

73. Fairclough translates: "Hence it is that, glad with some strange joy, they cherish nest and nestlings; hence they deftly mould fresh wax and fashion the gluey honey." Virgil, *Georgics*, 223.

74. Dryden, *Georgics*, 4.226–27, 230–31.

75. Ibid., 4.259–60, 270.

76. Ibid., 4.20.

77. Philips did not share that particular objection; he praises English-grown grain as the source of yet another quintessentially English drink, ale. See *Cyder*, 1.551–58.

78. Ibid., 1.422–23.
79. Ibid., 1.427, 434–36.

3. Pope and the Usufructuary Ethics of the "Use of Riches"

1. Pope, *Epistle to Burlington*, 89–92. This and all subsequent quotations of Pope's poetry come from *Poetry and Prose of Alexander Pope*.
2. Ibid., 93–98.
3. For a particularly thorough and insightful consideration of this topic, see Gibson, "Three Principles of Renaissance Architectural Theory."
4. See Mack, *The Garden and the City*, and *Alexander Pope: A Life*; Hammond, *Pope and Bolingbroke*; Erskine-Hill, "Pope and the Question of Jacobite Vision"; and Rogers, *A Political Biography of Alexander Pope*.
5. Mack, *The Garden and the City*, 96.
6. Hammond, *Pope and Bolingbroke*, 165 and 164.
7. Maresca, *Pope's Horatian Poems*, 143–44.
8. Ibid., 144.
9. Ibid., 130.
10. *The Bible: Authorized King James Version*.
11. Fleury was probably best known for *Les moeurs'* sustained defense of the honor and nobility of agrarian life in ancient societies (Greek and Roman as well as Jewish) against the degenerating effects of urban idleness and disconnection from the earth, which he relates to the virtue of ancient society versus modern. The book was first published in France in 1681, and it was published in English in 1683 as *The Manners of the Israelites, in Three Parts*, by an unnamed translator. A new English translation appeared in 1756, followed by yet another in 1786. Fleury's condemnation of the "gothic" barbarity of the aristocratic love of hunting in *Les moeurs* was referred to by Knightley Chetwood in his preface to Dryden's translation of Virgil's *Eclogues*, and later by Pope in his essay on cruelty to animals in *The Guardian*. See Chetwood, "Preface to the *Pastorals*," in Dryden, *Works*, 5:38; and Pope, *The Guardian*, 234.
12. Fleury, *The Manners of the Israelites*, Cap. 2. The tension between seeing the world as a source of subsistence versus profit is also fundamental to the georgics of the eighteenth century. I will return to this idea in the fourth chapter.
13. Ibid., 60, my emphasis.
14. Dodd, *Commentary on the books of the Old and New Testament*, n.p. According to his biographer, Dodd "published a *Commentary of the Bible* in monthly parts from 1764 and became almost solely responsible for the Christian Magazine (1760–67)" during that same period. Dodd's monthly installments were collected and published as a book in 1770. An

advertisement appears in the *Gazetteer and New Daily Advertiser* (London), on Wednesday, April 24, 1765, for the "publication by subscription" of Dodd's *Commentaries*. Dodd was a popular preacher in London in the 1760s and was tutor to Philip Stanhope, Earl of Chesterfield from 1765 to 1771. His career ended in ignominy, however, when he was found guilty of forging a bill of exchange on the Earl of Chesterfield. He was sentenced to death, and despite a public outcry that included vocal support from Samuel Johnson (who believed in the cause though he did not personally like Dodd), Lord Mansfield upheld the sentence. Dodd was hanged for capital forgery on June 27, 1777. See Rawlings, "Dodd, William (1729–1777)."

15. Pope, *Satire* 2.2, 161–67.
16. Pope, *Epistle* 2.2, 246–51.
17. Maresca, *Pope's Horatian Poems*, 118.
18. Ibid., 219.
19. Pope, *Satire* 2.2, 117–18.
20. Stack, *Pope and Horace*, 68. Incidentally, Stack also remarks on the way that "'dominium,' owning property, gives way to 'usufructus,' the right only of using and enjoying it" in *Satire* 2.2; and like Maresca, points toward Lucretius's lines on displacement as a possible source (70). We now know, of course, that Lucretius would have been one of dozens of popular sources espousing that idea during Pope's lifetime.
21. Pope, *Satire* 2.2, 105–10.
22. Pope, *Epistle to Burlington*, 43.
23. Ibid., 57.
24. Pope, "Argument" to *Epistle to Burlington*, 189.
25. Wasserman, *Pope's Epistle to Bathurst*, 37–38. A portion of Wasserman's analysis of *Bathurst* turns on his argument that *Bathurst's* moral argument is more thorough and sustained than *Burlington's*. In his account, *Burlington* is primarily about aesthetics, not ethics.
26. Gibson, "Three Principles of Renaissance Architectural Theory," 489.
27. Engell, "Wealth and Words," 443, my emphasis. Tom Jones takes a similar angle in "Pope's *Epistle to Bathurst* and the Meaning of Finance."
28. In addition to Mack, see Brown, *Alexander Pope*, 94–127; and Kelsall, *The Great Good Place*, 59–88, and "Landscapes and Estates."
29. Chaucer, "The Parson's Tale," 313.
30. Aubrey Williams, "A Hell for 'Ears Polite,'" 486.
31. To wit, Robert South (1634–1716), from a collection of sermons first published in 1694: "No Man holds the Abundance of Wealth, Power, and Honour, that Heaven has blessed him with, as a *Proprietor*, but as a *Steward*, as the Trustee of Providence to use and dispense it for the Good of those whom he converses with. . . . God bids a great and rich Person rise and shine, as he bids the Sun; that is, not for himself, but for the

Necessities of the World." In this short passage, we find both displacement and mediality, as South insists not only on God's persistent ownership but also on human beings' (and particularly landlords') role as the conduits through whom the gifts of God's creation pass to others further down the social-ontological hierarchy. Robert South, *Twelve Sermons*, 4:74–75; quoted in A. Williams, "A Hell for 'Ears Polite,'" 486. A "sixth edition" of South's *Twelve Sermons* dated 1727 is the latest to appear in *Eighteenth Century Collections Online.* The chaplain to Clarendon, Duke of York, and Charles II, South was a "court preacher" and very much in the Anglican mainstream of preaching and controversy in post-Restoration England.

32. Pope, *Epistle to Burlington*, 179–90.
33. Ibid., 177.
34. Ibid., 65–66.
35. Pope, *Essay on Man*, 3.21–24.
36. Ibid., 3.7–8, and "The Design," 120.
37. Smith, "Political Individuals and Providential Nature," 619; see also C. W. Smith, *Empiricist Devotions*.
38. Allestree, *The Gentleman's Calling*, 294–95.
39. Pope, *Epistle to Burlington*, 68–69.
40. Ibid., 66.
41. Ibid., 79–82.
42. Pope, *Windsor-Forest*, 14.
43. Espaliers, as Aubrey Williams notes, are "fruit trees trained upon a framework of stakes." Pope, *Epistle to Burlington*, note to line 80.
44. Ibid., 87–88.
45. Engell, "Wealth and Words," 443.
46. Kelsall, "Landscapes and Estates," 170.
47. Mack, *Alexander Pope*, 515.
48. Pope, *Epistle to Bathurst*, 229–34.
49. Swift, "On Mutual Subjection," in *Four Sermons*, 3. For a full discussion of the relevance of this sermon to the usufructuary ethos and the socioenvironmental ideology of the landlord, see above, chapter 2.
50. Pope, *Epistle to Burlington*, 16.
51. Pope, *Satire* 2.2, 111–18.
52. Adams, "The Barren Tree," in *Five Sermons*, 27. For the full discussion of this passage's significance, see chapter 2.
53. C. W. Smith, *Empiricist Devotions*, 5.
54. I am indebted to Maynard Mack and Frank Stack for calling my attention to the disparity between Horace and Pope, and to Stack for the English translation of Horace's lines. See Mack, *Alexander Pope*, 591–92; and Stack, *Pope and Horace*, 67–68.
55. Pope, *Epistle to Bathurst*, 253–58.

56. Ibid., 262.
57. Ibid., 259.
58. Ibid., 219–20.
59. Fager, *Land Tenure and the Biblical Jubilee*, 112–13.
60. I am indebted throughout this section to Dwight Codr's nuanced complication of this history of "finance" in the early eighteenth century, particularly his theorization of the ethics of "anti-finance" and its resistance to the prioritization of profit and certainty. See Codr, *Raving at Usurers*.
61. Fuchs, *Reading Pope's Imitations of Horace*, 83–84.
62. Pope, *Satire* 2.2, 133 and 136; my emphasis.
63. Ibid., 165–66.
64. Discussions of this issue that focus on *Burlington* or *Bathurst* in particular include Wasserman, *Pope's Epistle to Bathurst*; Engell, "Wealth and Words"; Tom Jones, "Pope's *Epistle to Bathurst*"; Noggle, "Taste and Temporality." For discussions of Pope's general attitude toward the financial revolution, see Kramnick, *Bolingbroke and His Circle*, esp. 217–23; Erskine-Hill, "Pope and the Financial Revolution"; and Mack, *Alexander Pope*.
65. See, e.g., Christine Garrard, *The Patriot Opposition to Walpole*; and Erskine-Hill, "Pope and the Poetry of Opposition."
66. Engell, "Wealth and Words," 438, my emphasis.
67. Mack, *Alexander Pope*, 513.
68. Pope, *Epistle to Burlington*, 98.
69. Ibid., 93.
70. Ibid., 15–16, my emphasis.
71. *The Landlord's Law*, 276.
72. Pope, *Epistle to Burlington*, 119, 163.
73. Gibson, "Three Principles of Renaissance Architectural Theory," 489.
74. Pope, *Epistle to Burlington*, 108.
75. Ibid., 119–20.
76. Ibid., 124, 126.
77. Ibid., 99–100.
78. Noggle, "Taste and Temporality," 122. Noggle offers a sensitive analysis of the role time plays in Pope's landscapes in *Burlington*, but I differ with him on the definitions of "sense" and "taste."
79. Pocock, *Virtue, Commerce, and History*, 114.
80. Ibid., 103.
81. Pope, *Epistle to Burlington*, 169–72.
82. Ibid., 173–76.
83. Wasserman, *Pope's Epistle to Bathurst*, 16–17.
84. Pope, *Epistle to Bathurst*, 163–64.
85. Wasserman, *Pope's Epistle to Bathurst*, 27.
86. Pope, *Epistle to Bathurst*, 174, 177–78.

87. Mack, *Alexander Pope*, 515.
88. Pope, *Epistle to Bathurst*, 193–98.
89. Ibid., 188.
90. On the invocation of Christian teachings on charity in the Cotta passage, see Wasserman, *Pope's Epistle to Bathurst*, 17–18.
91. Pope, *Epistle to Bathurst*, 219–20, my emphasis.
92. C. W. Smith, *Empiricist Devotions*, 131.
93. Wasserman, *Pope's Epistle to Bathurst*, 16–17.
94. Pope, *Epistle to Bathurst*, 206, 204.
95. Ibid., 209–10.
96. Ibid., 217–18.
97. Probably the most prominent versions of this argument are John Barrell and Harriet Guest, "On the Use of Contradiction"; and Laura Brown, *Alexander Pope*, chap. 3, "The Ideology of Neo-Classical Aesthetics: *Epistles to Several Persons*."
98. Pope, *Epistle to Bathurst*, 332.
99. Wasserman, *Pope's Epistle to Bathurst*, 44. Wasserman's full discussion of the figure of Balaam incorporates the biblical Balaam with Popean Balaam's Whig, Dissenting, and capitalistic values (44–55). Engell further ties *Bathurst's* "Sibylline leaves," a reference to the books of Augustan Roman religious rites, to the "leaves" of paper credit "essential to Britain's new religion, its ministers, its Augustus" ("Wealth and Words," 445).
100. Pope, *Epistle to Bathurst*, 390, 392, 396, 400.
101. Christine Garrard, *The Patriot Opposition to Walpole*.
102. Pope, *Epistle to Burlington*, 186–87, 197, 204.
103. Pope, *Windsor-Forest*, 386, 398–400.
104. Feingold, *Nature and Society*, 36.

4. Monocultures, Georgics, and the Transformation of the Usufructuary Ethos

1. Recent georgic scholarship owes deep debts to seminal works by Dwight L. Durling (*Georgic Tradition in English Poetry*), L. P. Wilkinson (*The Georgics of Virgil*), John Chalker (*The English Georgic*), and Anthony Low (*The Georgic Revolution*). More recently, Frans De Bruyn has done important work connecting georgic's popularity to the period's fascination with agricultural science. See De Bruyn, "Reading Virgil's *Georgics* as a Scientific Text," and "Eighteenth-Century Editions of Virgil's *Georgics*."
2. The foundational texts for such readings of the georgic are Raymond Williams's *The Country and the City*; and John Barrell's *The Dark Side of Landscape*, and *English Literature in History*. More recently, Karen O'Brien ("Imperial Georgic, 1660–1789") has tied the rise and decline of georgic as

a popular literary genre in eighteenth-century Britain to its relationship to empire, especially as a mode of description of natural-economic activity that justifies and valorizes British imperial activities. See also Irvine, "Labor and Commerce in Locke."

3. Probably the earliest essay to make an explicit claim for georgic's value as an "ecological" genre was Donna Landry's "Georgic Ecology." David Fairer's pro-georgic polemic, "'Where fuming trees refresh,'" made a broad claim for the vital insights of eighteenth-century georgic for twenty-first-century environmentalism, particularly in light of the genre's insistence on cooperation with nature over mastery, which countered the widely accepted narrative of the eighteenth century as anti-ecological common among eco-critics in the 1990s and early 2000s. See also Drew, "'Iron War' as 'Daily Care'"; and C. W. Smith, *Empiricist Devotions*, 173–210.

4. Fairer, "'Where fuming trees refresh,'" 207.

5. Loar, "Georgic Assemblies," 242.

6. Genovese, "An Organic Commerce," 198. While not disagreeing with Genovese's basic point about the ways midcentury georgics attempted to reconcile the changing agricultural economy to older, nostalgic social values, this chapter takes a very different view of the meanings and outcomes of those attempts.

7. For another discussion of this connection, see Drew, "'Tis Prudence to prevent.'"

8. For a recent analysis of the "fixation on property" in Grainger, particularly as it relates to the depiction and defense of slavery, see Schweiger, "Grainger's West Indian Planter."

9. Feingold, *Nature and Society*, 9, 93.

10. Ibid., 90.

11. Irvine, "Labor and Commerce in Locke," 985.

12. O'Brien, "Imperial Georgic, 1660–1789," 172–73. For another analysis of the influence of slavery on North American georgics in the eighteenth century, see Timothy Sweet, *American Georgics*.

13. Kaul, *Poems of Nation, Anthems of Empire*, 223.

14. See in particular Genovese, "An Organic Commerce"; and Loar, "Georgic Assemblies."

15. Kaul, *Poems of Nation, Anthems of Empire*, 203.

16. Crawford, "English Georgic and British Nationhood," 135.

17. Kaul, *Poems of Nation, Anthems of Empire*, 224.

18. Dyer, *The Fleece*, 1.357–58. Further citations to this work are in the text.

19. Crawford, "English Georgic and British Nationhood," 135–36.

20. Fairer, "'Where fuming trees refresh,'" 204–5.

21. Moore, "Sugar and the Expansion of the Early Modern World Economy," 416.

22. Ibid., 417.
23. Grainger, *The Sugar-Cane*, 1.559; further citations to this work are in the text. John Gilmore identified the definition of the "contagious blast" in *The Poetics of Empire*, 253–54.
24. On deforestation and soil erosion on English island colonies in the seventeenth and eighteenth centuries, see Grove, *Green Imperialism*, particularly chaps. 3, 4, and 6.
25. Virgil, *Georgics*, in *Virgil I*, 227, 229.
26. Gilmore, *The Poetics of Empire*, 238.
27. Dyer, *Fleece*, 1.635, 645.
28. For an in-depth discussion of the centrality of the concept of property to *Sugar-Cane* and particularly its ramifications for slavery and the slave trade, see Schweiger, "Grainger's West Indian Planter."
29. Dryden, *Georgics*, 1.264–71.
30. Virgil, *Georgics*, in *Virgil I*, 111; the translation is of Virgil, 1.185–86.
31. Doody, *The Daring Muse*, 111.
32. *OED Online*, s.v. "abide," my emphasis.
33. Dryden, *Georgics*, 1.179–80. Dryden tells the story of Jove deciding to forbid "Plenty to be bought with Ease" in 1.184–206.
34. Ibid., 1.141–42.
35. On *Sugar-Cane* and its problems with eighteenth-century poetic decorum, see in particular Irlam, "'Wish You Were Here.'"
36. Review of *The Sugar-Cane*, in *The Critical Review* 18 (1764): 271. The review was published anonymously, but has been confidently attributed to Johnson.
37. Sugar cane is native to South Asia and was brought to the Americas by Europeans starting around the sixteenth century. Grainger acknowledges sugar cane's Asian origins in a lengthy footnote in book I, but muddles the issue somewhat for the Caribbean islands by claiming that whether or not "the Cane is a native of either the Great or Lesser Antilles cannot now be determined, for their discoverers were so wholly employed in searching after imaginary gold-mines, that they took little or no notice of the natural productions." Grainger, *The Sugar-Cane*, note to 1.22.
38. Ferguson, *An Essay on the History of Civil Society*, 122.

Conclusion

1. Orr, *Dangerous Years*, 129.
2. Dragonetti, *A treatise on virtues and rewards*, 71.
3. Holthaus, "The Kids Suing the Government."
4. See, for instance, Blumm and Wood, chap. 1, "Introduction," in *The Public Trust Doctrine*, 1–56.

5. Holthaus, "The Kids Suing the Government."
6. See, e.g., Alaimo, "Sustainable This, Sustainable That." For a more detailed history of debate over the term and definition of "sustainability" in environmental discourse, and a defense of the concept, see Philippon, "Sustainability and the Humanities," 166–68.
7. Pope Francis, "Laudato si."
8. Marsh, *Man and Nature,* 35.
9. Berry, *What Are People For?,* 99–100.
10. Locher and Fressoz, "Modernity's Frail Climate," 598.

BIBLIOGRAPHY

Primary Sources

Adams, Thomas. *Five Sermons Preached upon Sundry Especiall Occasions.* . . . London, 1626. *Early English Books Online.*

Addison, Joseph, and Richard Steele. *The Spectator.* 1711–12. Vol. 7. London, 1753. *Eighteenth Century Collections Online.*

Allestree, Richard. *The art of contentment, by the author of the Whole duty of man, &c.* 1675. Oxford, 1705. *Eighteenth Century Collections Online.*

———. *The Gentleman's Calling.* In *The Works of the Learned and Pious Author of the Whole Duty of Man.* 393–453.

———. *The Whole Duty of Man, Laid Down for the Use of All, but especially the Meanest Reader.* 1659. London, 1703. *Eighteenth Century Collections Online.*

———. *The Works of the Learned and Pious Author of The Whole Duty of Man.* Oxford, 1704. *Eighteenth Century Collections Online.*

Ames, William. *Conscience with the power and cases thereof, divided into five bookes written by . . . William Ames . . . ; translated out of Latine into English for more publique benefit.* London, 1643.

Aquinas, Thomas. *Basic Writings of Saint Thomas Aquinas.* Ed. Anton C. Pegis. Vol. 2. New York: Random House, 1945.

Ayliffe, John. *A New Pandect of Roman Civil Law, As Anciently Established in that Empire; and Now Received and Practiced in Most European Nations.* Vol. 1. London, 1734. *Eighteenth Century Collections Online.*

Bacon, Francis. *Law Tracts.* London, 1737. *Eighteenth-Century Collections Online.*

The Bible: Authorized King James Version. Oxford World's Classics. Oxford: Oxford University Press, 1997.

Blackstone, William, Sir. *An Analysis of the Laws of England.* Oxford, 1756. *Eighteenth Century Collections Online.*

———. *Commentaries on the Laws of England.* Vol. 2. Oxford, 1766. *Eighteenth Century Collections Online.*

Burlamaqui, J. J. *The Principles of Politic Law. Being a Sequel to the Principles of Natural Law.* Trans. Mr. Nugent. London: J. Nourse, 1752. *Eighteenth Century Collections Online.*

Chaucer, Geoffrey. "The Parson's Tale." In *The Riverside Chaucer,* ed. Larry D. Benson. 3rd ed. New York: Houghton Mifflin, 1987. 287–327.

Cowell, John. *Institutes of the Law of England.* Trans. W. G. London. 1651. *Early English Books Online.*

Cowper, William. *The Poems of William Cowper.* Ed. John D. Baird and Charles Ryskamp. 3 vols. New York: Oxford University Press, 1995.

Dalrymple, John. *An Essay Towards a General History of Feudal Property in England.* 2nd ed. London: A. Millar, 1758. *Eighteenth Century Collections Online.*

The Digest of Justinian. English translation ed. Alan Watson. Philadelphia: University of Pennsylvania Press, 1985.

Dodd, William. *A commentary on the books of the Old and New Testament.* Vol. 1. London, 1770. *Eighteenth Century Collections Online.*

Domat, Jean. *Les Lois Civiles dans leur Ordre Naturel.* New ed. Vol. 1. The Hague: Chez Adrian Moetjens, 1703. *Google Books.* Trans. William Strahan as *The Civil Law in its Natural Order: Together with the Publick Law.* 2 vols. (London, 1722).

Dragonetti, Giacinto, marchese. *A treatise on virtues and rewards.* Trans. Henry Fuseli. London, 1769. *Eighteenth Century Collections Online.*

Dryden, John. *Georgics.* In *The Works of John Dryden.* Vol. 5. Ed. William Frost and Vinton A. Dearing. Berkeley: University of California Press, 1987.

Dyer, John. *The Fleece: A Poem. In Four Books.* London: R. and J. Dodsley, 1757. *Eighteenth Century Collections Online.*

Dykes, Oswald. *Moral reflexions upon select British proverbs: familiarly accommodated to the humour and manners of the present age, after the Method of Sir Roger L'Estrange.* London, 1708. *Eighteenth Century Collections Online.*

———. *The Royal Marriage. King Lemuel's Lesson on Chastity, Temperance, Charity, Justice, Education, Industry, Frugality, Religion, Marriage, &c. Practically Paraphras'd, with Remarks, Moral and Religious, upon the Virtues and Vices of Wedlock.* London, 1722. *Eighteenth Century Collections Online.*

Evelyn, John. *Silva. Or, a Discourse of Forest-Trees, and the Propagation of Timber in His Majesty's Dominions.* 4th ed. London, 1706. *Eighteenth Century Collections Online.*

Ferguson, Adam. *An Essay on the History of Civil Society.* Ed. Duncan Forbes. Edinburgh: Edinburgh University Press, 1966.

Finch, Anne. *The Poems of Anne, Countess of Winchilsea.* Ed. Myra Reynolds. Chicago: University of Chicago Press, 1903.

Flavel, John. *Husbandry Spiritualiz'd: Or, the Heavenly Use of Earthly Things. . . .* 8th ed. London, 1714. *Eighteenth Century Collections Online.*

Fleury, Abbé Claude. *The Manners of the Israelites, in Three Parts.* London, 1683. *Early English Books Online.*

———. *Manners of the Israelites. Wherein Is Seen the Model of a Plain and Honest Policy for the Government of States, and Reformation of Manners.* Trans. Charles Cordell. Newcastle, 1786. *Eighteenth Century Collections Online.*

———. *A Short History of the Israelites. With an Account of their Manners, Customs, Laws, Polity and Religion.* Trans. Ellis Farneworth. London, 1756. *Eighteenth Century Collections Online.*

Gilbert, Sir Jeffrey. *The Law of Uses and Trusts.* London, 1734. *Eighteenth-Century Collections Online.*

Grace, John. *A Sermon Preached in the Parish-Church of Lisburn, on the 21st of June, 1749.* . . . Dublin, 1750. *Eighteenth Century Collections Online.*

Grainger, James. *The Sugar-Cane: A Poem in Four Books.* London: R. and J. Dodsley, 1764. *Eighteenth Century Collections Online.*

Grotius, Hugo. *The Rights of War and Peace.* 1625. Ed. Richard Tuck. Trans. John Morrice. 1738. 3 vols. Indianapolis: Liberty Fund, 2005.

The Guardian. Ed. John Calhoun Stephens. Lexington: University Press of Kentucky, 1982.

Guénée, Antoine. *Letters of Certain Jews to Monsieur de Voltaire.* Trans. Philip LeFanu. Vol. 2. Dublin, 1777. *Eighteenth-Century Collections Online.*

Hale, Sir Matthew. *Contemplations Moral and Divine.* 5th ed. 2 vols. Edinburgh, 1792. *Eighteenth Century Collections Online.*

———. *An Epitome of Judge Hale's Contemplations, in His Account of the Good Steward.* Bristol, 1766. *Eighteenth Century Collections Online.*

———. *The great audit, or good steward. Being some necessary and important considerations, to be considered of by all sorts of people.* Liondon, 1775. *Eighteenth Century Collections Online.*

———. *The Primitive Origination of Mankind.* London, 1677. *Eighteenth Century Collections Online.*

———. *Some necessary and important considerations, directed to all sorts of people, taken out of the writings of that late worthy and renowned judge, Sir Matthew Hale.* 12th ed. Woodbridge, 1759. *Eighteenth Century Collections Online.*

Hales, John. *The Works of the Ever Memorable Mr. John Hales of Eaton.* Vol. 2. Glasgow, 1765. *Eighteenth Century Collections Online.*

Hallifax, Samuel. *An analysis of the Roman civil law, compared with the laws of England.* Cambridge, 1774. *Eighteenth Century Collections Online.*

Horneck, Anthony. *The first fruites of reason, or, A discourse shewing the necessity of applying ourselves betimes to the serious practice of religion.* London, 1686. *Early English Books Online.*

Howe, John. *Self-Dedication Discoursed in the Anniversary Thanksgiving of a Person of Honour for a Great Deliverance.* London, 1682. *Early English Books Online.*

Jacob, Giles. *A Treatise of Laws: Or, a General Introduction to the Common, Civil and Canon Law.* London, 1721. *Eighteenth Century Collections Online.*

Jefferson, Thomas. *Political Writings.* Ed. Joyce Oldham Appleby and Terence Ball. New York: Cambridge University Press, 1999.

Johnson, Samuel. *Dictionary of the English Language.* 2nd ed. Vol. 2. London, 1755–56.

———. Review of *The Sugar-Cane,* by James Grainger. *The Critical Review,* vol. 18. London, 1764.

Kidder, Richard. "The Life of the Reverend Anthony Horneck, D.D." In *Several Sermons Upon the Fifth of St. Matthew*, by Anthony Horneck. 2nd ed. Vol. 1. London, 1706. *Eighteenth-Century Collections Online*.

Knox, Vicesimus. *Essays Moral and Literary*. 6th ed. London, 1785. *Eighteenth Century Collections Online*.

The Landlord's Law; or, the Law concerning Landlords, Tenants, and Farmers. 6th ed. London, 1720. *Eighteenth-Century Collections Online*.

Locke, John. *An Essay Concerning Human Understanding*. 1689. Ed. Peter H. Nidditch. New York: Oxford University Press, 1975.

———. *Essays on the Law of Nature*. Ed. and trans. W. von Leyden. Oxford: Oxford University Press, 1954.

———. *Two Treatises of Government*. Ed. Peter Laslett. New York: Cambridge University Press, 1960.

Philips, John. *The Cyder. A Poem in Two Books*. London: Jacob Tonson, 1708. *Eighteenth Century Collections Online*.

———. *Cyder: A Poem in Two Books*. Ed. John Goodridge and J. C. Pellicer. Cheltenham, England: The Cyder Press, 2001.

Physiologues, Philotheos [Thomas Tryon]. *The Country-Man's Companion: or, A New Method of Ordering Horses and Sheep.* . . . London, [1684]. *Early English Books Online*.

Place, Conyers. *An Essay Towards the Vindication of the Visible Creation.* . . . *Book II*. London, 1729. *Eighteenth Century Collections Online*.

Pope, Alexander. *Poetry and Prose of Alexander Pope*. Ed. Aubrey Williams. Boston: Houghton Mifflin, 1969.

Prideaux, Humphrey. *The Original and Right of Tithes, for the Maintenance of the Ministry in a Christian Church Truly Stated.* . . . Norwich, 1710. *Eighteenth-Century Collections Online*.

Pufendorf, Samuel. *The Law of Nature and Nations: or, a General System.* . . . Trans. Basil Kennet. 5th ed. London, 1749. *Eighteenth Century Collections Online*.

———. *The whole duty of man according to the law of nature*. Ed. Ian Hunter and David Saunders. Indianapolis: Liberty Fund, 2003.

Richardson, Samuel. *Pamela: Or, Virtue Rewarded*. Ed. Thomas Keymer and Alice Wakely. Oxford World's Classics. New York: Oxford University Press, 2001.

Rose, Hugh. *Meditations on several interesting subjects*. Edinburgh, 1762. *Eighteenth-Century Collections Online*.

South, Robert. *Twelve Sermons Preached at Several Times, and upon Several Occasions*. 4th ed. 6 vols. London, 1727.

Stillingfleet, Edward. *Irenicum, a Weapon-Salve for the Churches Wounds, or the Divine Right of Particular Forms of Church-Government.* . . . London, 1662. *Early English Books Online*.

Strahan, William. Translator's preface to *The Civil Law in its Natural Order: Together with the Publick Law*, by Jean Domat. Trans. William Strahan. 2 vols. London, 1722. *Eighteenth Century Collections Online.*

Sturm, C. C. *Reflections for Every Day of the Year, on the Works of God, and of His Providence Throughout Nature.* 7th ed. 3 vols. Edinburgh: N. R. Cheyne, St. Andrew's St., 1800.

Sullivan, Francis Stoughton. *An Historical Treatise on the Feudal Law, and the Constitution and Laws of England; with a Commentary on Magna Charta, and necessary Illustrations of many of the English Statutes, in a Course of Lectures, read in the University of Dublin.* Dublin, 1752. *Eighteenth Century Collections Online.*

Swift, Jonathan. *The Beauties of Swift, or, the Favourite Offspring of Wit and Genius.* 2nd ed. London, 1782. Eighteenth Century Collections Online.

———. *Four Sermons.* Dublin, 1760. Eighteenth Century Collections Online.

Towerson, Gabriel. *Explication of the Decalogue.* London, 1676. Eighteenth Century Collections Online.

Tryon, Thomas. See Physiologues, Philotheos.

Virgil. *The Georgics.* Trans. L. P. Wilkinson. New York: Penguin, 1982.

———. *The Georgics: A Poem of the Land.* Trans. Kimberley Johnson. New York: Penguin, 2009.

———. *Virgil I: Eclogues; Georgics; Aeneid I–VI.* Trans. H. Rushton Fairclough. Loeb Classical Library 63. 1935. Rev. G. P. Goold. Cambridge, MA: Harvard University Press, 1999.

Wollaston, William. *The religion of nature delineated.* London, 1722. *Eighteenth Century Collections Online.*

Wood, Thomas. Some Thoughts Concerning the Study of the Laws of England in the Two Universities. London, 1708. *Eighteenth Century Collections Online.*

———. *A New Institute of the Imperial or Civil Law. With Notes, Shewing in some Principal Cases, amongst other Observations, how the Canon Law, the Laws of England and Customs of other Nations differ from it.* 4th ed. 4 vols. London: J. and J. Knapton, et al., 1730. *Eighteenth Century Collections Online.*

Wycherley, William. *Miscellany Poems: as Satyrs, Epistles, Love-Verses, Songs, Sonnets, &c.* London, 1704. *Eighteenth-Century Collections Online.*

Secondary Sources

Alaimo, Stacy. "Sustainable This, Sustainable That: New Materialisms, Posthumanism, and Unknown Futures." *PMLA* 127.3 (2012): 558–64.

Allen, David Elliston. *The Naturalist in Britain: A Social History.* London: Allen Lane, 1976.

Anderson, David. "Thomson, Natural Evil, and the Eighteenth-Century Sublime." *Studies on Voltaire and the Eighteenth Century* 245 (1986): 489–99.

Arthos, John. *The Language of Natural Description in Eighteenth-Century Poetry.* Ann Arbor: University of Michigan Press, 1949.

Attfield, Robin. "Western Traditions and Environmental Ethics." In *Environmental Philosophy: A Collection of Readings,* ed. Robert Elliot and Arran Gare. University Park: Pennsylvania State University Press, 1983.

Auerbach, Bruce Edward. *Unto the Thousandth Generation: Conceptualizing Intergenerational Justice.* New York: Peter Lang, 1995.

Ayers, Philip. "Pope's Epistle to Burlington: The Vitruvian Analogies." *Studies in English Literature 1500–1900* 30.3 (1990): 429–44.

Barad, Karen. *Meeting the Universe Halfway: Quantum Physics and the Entanglement of Matter and Meaning.* Durham: Duke University Press, 2007.

Barrell, John. *The Dark Side of Landscape: The Rural Poor in English Paining, 1730–1840.* Cambridge: Cambridge University Press, 1980.

———. *English Literature in History, 1730–80: An Equal, Wide Survey.* New York: St. Martin's Press, 1983.

———. *The Idea of Landscape and the Sense of Place, 1730–1840: An Approach to the Poetry of John Clare.* New York: Cambridge University Press, 1972.

Barrell, John, and Harriet Guest. "On the Use of Contradiction: Economics and Morality in the Eighteenth-Century Long Poem." In *The New Eighteenth Century,* ed. Felicity Nussbaum and Laura Brown. London: Routledge, 1987. 121–43.

———. "Thomson in the 1790s." In *James Thomson: Essays for the Tercentenary,* ed. Richard Terry. Liverpool: Liverpool University Press, 2000. 217–46.

Bate, Jonathan. Forward to *The Green Studies Reader: From Romanticism to Ecocriticism,* ed. Laurence Coupe. New York: Routledge, 2000. xvii.

———. *The Song of the Earth.* London: Picador, 2000.

Batstone, William. "Virgilian Didaxis: Value and Meaning in the *Georgics.*" In *The Cambridge Companion to Virgil,* ed. Charles Martindale. New York: Cambridge University Press, 1997. 125–44.

Bergthaller, Hannes. "Introduction: Ecocriticism and Environmental History." *ISLE: Interdisciplinary Studies in Literature and Environment* 22.1 (2015): 5–8.

———. "'No More Eternal than the Hills of the Poets': On Rachel Carson, Environmentalism, and the Paradox of Nature." *ISLE: Interdisciplinary Studies in Literature and Environment* 22.1 (2015): 9–26.

Berry, Wendell. "Poetry and Place." In *Standing by Words.* San Francisco: North Point Press, 1983. 92–199.

———. *What Are People For?* New York: North Point Press, 1990.

Blau, Sheridan D. "Pope's 'Chain of being' and the Modern Ecological Vision." *CEA Critic: An Official Journal of the College English Association* 33.2 (1971): 20–22.

Blumm, Michael C., and Mary C. Wood. *The Public Trust Doctrine in Environmental and Natural Resources Law.* Durham, NC: Carolina Academic Press, 2013.

Bonneuil, Christophe, and Jean-Baptiste Fressoz. *The Shock of the Anthropocene.* 2013. Trans. David Fernbach. New York: Verso, 2016.

Bowerbank, Sylvia. *Speaking for Nature: Women and Ecologies of Early Modern England.* Baltimore: Johns Hopkins University Press, 2004.

Braidotti, Rosi. *Transpositions.* Malden, MA: Polity, 2006.

Brock, Gillian. "Future Generations, Natural Resources, and Property Rights." *Ethics and the Environment* 3.2 (1998): 119–30.

Brown, Laura. *Alexander Pope.* Oxford: Blackwell, 1985.

Buckle, Stephen. *Natural Law and the Theory of Property: Grotius to Hume.* New York: Oxford University Press, 1991.

Buell, Lawrence. *The Future of Environmental Criticism: Environmental Crisis and Literary Imagination.* Malden, MA: Blackwell, 2005.

Caldwell, Tanya. *Time to Begin Anew: Dryden's Georgics and Aeneis.* Cranbury, NJ: Associated University Presses, 2000.

———. *Virgil Made English: The Decline of Classical Authority.* New York: Palgrave Macmillan, 2008.

Campbell-Culver, Maggie. *A Passion for Trees: The Legacy of John Evelyn.* London: Eden Project Books, 2006.

Caradonna, Jeremy L. *Sustainability: A History.* Oxford: Oxford University Press, 2014.

Carey, Daniel, and Lynn Festa, eds. *The Postcolonial Enlightenment: Eighteenth-Century Colonialism and Postcolonial Theory.* New York: Oxford University Press, 2009.

Casid, Jill. *Sowing Empire: Landscape and Colonization.* Minneapolis: University of Minnesota Press, 2005.

Castellano, Katey. *The Ecology of British Romantic Conservatism, 1790–1837.* New York: Palgrave Macmillan, 2013.

Chakrabarty, Dipesh. "The Climate of History: Four Theses." *Critical Inquiry* 35.1 (2009): 197–222.

Chalker, John. *The English Georgic: A Study in the Development of a Form.* Baltimore: Johns Hopkins University Press, 1969.

———. "Thomson's *Seasons* and Virgil's *Georgics*: The Problems of Primitivism and Progress." *Studia Neophilologica: A Journal of Germanic and Romance Languages and Literature* 35 (1963): 41–56.

Chambers, Douglas. *The Planters of the English Landscape Garden: Botany, Trees, and the Georgics.* New Haven: Yale University Press, 1993.

Clark, Timothy. *The Cambridge Introduction to Literature and the Environment.* New York: Cambridge University Press, 2011.

———. "Some Climate Change Ironies: Deconstruction, Environmental Politics and the Closure of Ecocriticism." *Oxford Literary Review* 32.1 (2010): 131–49.

Codr, Dwight. *Raving at Usurers: Anti-Finance and the Ethics of Uncertainty in England, 1692–1750*. Charlottesville: University of Virginia Press, 2016.

Coleman, William. "Providence, Capitalism, and Environmental Degradation: English Apologetics in an Era of Economic Revolution." *Journal of the History of Ideas* 37.1 (1976): 27–44.

Crawford, Rachel. "English Georgic and British Nationhood." *ELH* 65.1 (1998): 123–58.

Cromartie, Alan. "Hale, Sir Mathew (1609–1676)." *Oxford Dictionary of National Biography*. Online ed. Ed. Lawrence Goldman. Oxford: Oxford University Press, 2004.

———. *Sir Matthew Hale, 1609–1676: Law, Religion, and Natural Philosophy*. New York: Cambridge University Press, 1995.

Cronon, William. "The Trouble with Wilderness; or, Getting Back to the Wrong Nature." In *Uncommon Ground: Rethinking the Human Place in Nature*, ed. William Cronon. New York: Norton, 1995. 69–90.

Crosby, Alfred. *Ecological Imperialism: The Biological Expansion of Europe, 900–1900*. Cambridge: Cambridge University Press, 1986.

Cutting-Gray, Joanne, and James E. Swearingen. "System, the Divided Mind, and the Essay on Man." *SEL: Studies in English Literature 1500–1900* 32.3 (1992): 479–94.

Dalzell, Alexander. *The Criticism of Didactic Poetry: Essays on Lucretius, Virgil, and Ovid*. Toronto: University of Toronto Press, 1996.

Darley, Gillian. *John Evelyn: Living for Ingenuity*. New Haven: Yale University Press, 2006.

Darwall, Stephen. *The British Moralists and the Internal 'Ought,' 1640–1740*. Cambridge: Cambridge University Press, 1995.

Davie, Donald. *The Language of Science and the Language of Literature, 1700–1740*. London: Sheed and Ward, 1963.

Daw, Carl P., Jr. "Swift and the Whole Duty of Man." *American Notes and Queries* 8 (1970): 86–87.

De Bruyn, Frans. "The Classical Silva and the Generic Development of Scientific Writing in Seventeenth-Century England." *New Literary History* 32.2 (2001): 347–73.

———. "Eighteenth-Century Editions of Virgil's *Georgics*: From Classical Poem to Agricultural Treatise." *Lumen: Selected Proceedings from the Canadian Society for Eighteenth-Century Studies* 24 (2005): 149–63.

———. "From Georgic Poetry to Statistics and Graphs: Eighteenth-Century Representations and the 'State' of British Society." *Yale Journal of Criticism* 17.1 (2004): 107–39.

———. "From Virgilian Georgic to Agricultural Science: An Instance in the Transvaluation of Literature in Eighteenth-Century Britain." In *Augustan*

Subjects: Essays in Honor of Martin C. Battestin, ed. Albert J. Rivero. Newark: University of Delaware Press, 1997. 47–67.

———. "Reading Virgil's Georgics as a Scientific Text: The Eighteenth-Century Debate between Jethro Tull and Stephen Switzer." *ELH* 71.3 (2004): 661–89.

DeLoughrey, Elizabeth M., Renée K. Gosson, and George B. Handley, eds. *Caribbean Literature and the Environment: Between Nature and Culture*. New World Studies. Charlottesville: University of Virginia Press, 2005.

Doody, Margaret Anne. *The Daring Muse: Augustan Poetry Reconsidered*. New York: Cambridge University Press, 1985.

———. "Insects, Vermin, and Horses: *Gulliver's Travels* and Virgil's *Georgics*." In *Augustan Studies*, ed. Douglas Lane Patey and Timothy Keegan. Newark: University of Delaware Press, 1985. 147–74.

———. *A Natural Passion: A Study of the Novels of Samuel Richardson*. Oxford: Oxford University Press, 1974.

———. "Richardson's Politics." *Eighteenth-Century Fiction* 2.2 (1990): 113–26.

Drew, Erin. "'Iron War' as 'Daily Care': Sustainability and the Dialectic of Care in Dryden's *Georgics*." *1650–1850: Ideas, Aesthetics, and Inquiries in the Early Modern Era* 22 (2015): 217–37.

———. "'Tis Prudence to prevent th' entire decay': Usufruct and Environmental Thought." *Eighteenth-Century Studies* 49.2 (2016): 195–210.

Drew, Erin, and John Sitter. "Ecocriticism and Eighteenth-Century English Studies." *Literature Compass* 8/5 (2011): 227–39.

Duckworth, Alistair M. *The Improvement of the Estate: A Study of Jane Austen's Novels*. Baltimore: Johns Hopkins Press, 1971.

Dukes, Paul. *Minutes to Midnight: History and the Anthropocene Era from 1763*. New York: Anthem Press, 2011.

Duncan, Myrl L. "Property as a Public Conversation, Not a Lockean Soliloquy: A Role for Intellectual and Legal History in Takings Analysis." *Environmental Law* 26 (1996): 1095–160.

Durling, Dwight L. *Georgic Tradition in English Poetry*. New York: Columbia University Press, 1935.

Egan, Jim. "The 'Long'd-for Aera' of an 'Other Race': Climate, Identity, and James Grainger's *The Sugar-Cane*." *Early American Literature* 38.2 (2003): 189–212.

Elliott, Robert. "Future Generations, Locke's Proviso and Libertarian Justice." *Journal of Applied Philosophy* 3.2 (1986): 217–27.

Engell, James. "Wealth and Words: Pope's *Epistle to Bathurst*." *Modern Philology* 85.4 (1988): 433–46.

Erskine-Hill, Howard. "Pope and the Financial Revolution." In *Writers and Their Backgrounds: Alexander Pope*, ed. Peter Dixon. Athens: Ohio University Press, 1972. 200–229.

———. "Pope and the Poetry of Opposition." In *The Cambridge Companion to Alexander Pope*, ed. Pat Rogers. Cambridge: Cambridge University Press, 2007. 134–49.

———. "Pope and the Question of Jacobite Vision." In *Poetry of Opposition and Revolution, Dryden to Wordsworth*. Oxford: Oxford University Press, 1996. 57–108.

———. "Pope on the Origins of Society." In *The Enduring Legacy: Alexander Pope Tercentenary Essays*, ed. G. S. Rousseau and Pat Rogers. Cambridge: Cambridge University Press, 1988. 79–93.

Evans, James E. "Fielding, *The Whole Duty of Man, Shamela,* and *Joseph Andrews.*" *Philological Quarterly* 61.2 (1982): 212–29.

Fager, Jeffrey A. *Land Tenure and the Biblical Jubilee: Uncovering Hebrew Ethics through the Sociology of Knowledge. Journal for the Study of the Old Testament,* Supplement series 155. Sheffield, England: Sheffield Academic Press, 1993.

Fairer, David. "'All manag'd for the best': Ecology and the Dynamics of Adaptation." In *Citizens of the World: Adapting in the Eighteenth Century*, ed. Kevin L. Cope and Samara Cahill. Transits: Literature, Thought, & Culture 1650–1850. Lewisburg, PA: Bucknell University Press, 2015. xxv–xlvii.

———. "A Caribbean Georgic: James Grainger's *The Sugar-Cane.*" *Kunapipi* 25.1 (2003): 21–28.

———. *English Poetry of the Eighteenth Century, 1700–1789.* Longman Literature in English. New York: Longman, 2003.

———. "'Where fuming trees refresh the thirsty air': The World of Eco-Georgic." *Studies in Eighteenth Century Culture* 40 (2011): 201–18.

Feingold, Richard. *Nature and Society: Later Eighteenth-Century Uses of the Pastoral and Georgic.* New Brunswick, NJ: Rutgers University Press, 1978.

Ferkiss, Victor. *Nature, Technology, and Society: Cultural Roots of the Current Environmental Crisis.* New York: New York University Press, 1993.

Field, David P. "Howe, John (1630–1705)." In *Oxford Dictionary of National Biography*, ed. H. C. G. Matthew and Brian Harrison. Oxford: Oxford University Press, 2004. Online ed. Ed. Lawrence Goldman.

Finkelstein, Andrea. *Harmony and the Balance: An Intellectual History of Seventeenth-Century English Economic Thought.* Ann Arbor: University of Michigan Press, 2000.

Fizer, Irene. "'A Passion for Dead Leaves': Animated Landscapes and Static Canvases in *Sense and Sensibility.*" *South Atlantic Review* 76.1 (2011): 53–72.

Foy, Anna. "The Convention of Georgic Circumlocution and the Proper Use of Human Dung in Samuel Martin's *Essay upon Plantership.*" *Eighteenth-Century Studies* 49.4 (2016): 475–506.

———. "Grainger and the 'Sordid Master': Plantocratic Alliance in *The Sugar-Cane* and Its Manuscript." *Review of English Studies*, new series, 68.286 (2017): 708–33.

Frier, Bruce W. *A Casebook on Roman Family Law*. Oxford: Oxford University Press, 2004.

Fuchs, Jacob. *Reading Pope's Imitations of Horace*. Lewisburg, PA: Bucknell University Press, 1989.

Fulford, Tim. *Landscape, Liberty and Authority: Poetry, Criticism and Politics from Thomson to Wordsworth*. New York: Cambridge University Press, 1996.

Garrard, Christine. *The Patriot Opposition to Walpole: Politics, Poetry, and National Myth, 1725–1742*. Oxford: Oxford University Press, 1994.

Garrard, Greg. *Ecocriticism*. 2nd ed. New York: Routledge, 2012.

Genovese, Michael. "An Organic Commerce: Sociable Selfhood in Eighteenth-Century Georgic." *Eighteenth-Century Studies* 46.2 (2013): 197–221.

Gibson, William A. "Three Principles of Renaissance Architectural Theory in Pope's *Epistle to Burlington*." *Studies in English Literature 1500–1900* 11.3 (1971): 487–505.

Gilcrest, David W. *Greening the Lyre: Environmental Poetry and Poetics*. Reno and Las Vegas: University of Nevada Press, 2002.

Gilmore, John. *The Poetics of Empire: A Study of James Grainger's* The Sugar-Cane. New Brunswick, NJ: Athlone Press, 2000.

Glacken, Clarence J. *Traces on the Rhodian Shore: Nature and Culture in Western Thought from Ancient Times to the End of the Eighteenth Century*. Berkeley: University of California Press, 1967.

Goodman, Kevis. *Georgic Modernity and British Romanticism: Poetry and the Meditation of History*. New York: Cambridge University Press, 2004.

Gough, J. W. *Fundamental Law in English Constitutional History*. Oxford: Clarendon Press, 1955.

Grant, Charlotte, ed. *Flora*. Vol. 4 of *Literature and Science, 1660–1834*, ed. Judith Hawley. London: Pickering & Chatto, 2003.

Grayling, A. C. *The Age of Genius*. New York: Bloomsbury, 2016.

Griffin, Dustin. "Redefining Georgic: Cowper's *Task*." *ELH* 57.4 (1990): 865–79.

Grober, Ulrich. *Sustainability: A Cultural History*. Trans. Ray Cunningham. Totnes, Devon: Green Books, 2012.

Grove, Richard H. *Green Imperialism: Colonial Expansion, Tropical Island Edens, and the Origins of Environmentalism, 1600–1860*. New York: Cambridge University Press, 1995.

Haakonssen, Knud. "Divine/Natural Law Theories in Ethics." In *The Cambridge History of Seventeenth-Century Philosophy*, ed. Daniel Garber and Michael Ayers. New York: Cambridge University Press, 1998. 2:1317–57.

———. *Natural Law and Moral Philosophy: From Grotius to the Scottish Enlightenment*. New York: Cambridge University Press, 1996.

Hammond, Brean. *Pope and Bolingbroke: A Study of Friendship and Influence*. New Haven: Yale University Press, 1985.

Hamrick, Wes. "Trees in Anne Finch's Jacobite Poems of Retreat." *SEL: Studies in English Literature 1500–1900* 53.3 (2013): 541–63.

Harding, Alan. *A Social History of English Law.* Baltimore: Penguin, 1966.

Hargrove, Eugene C. "Foundations of Wildlife Protection Attitudes." In *The Animal Rights/Environmental Ethics Debate: The Environmental Perspective,* ed. Eugene C. Hargrove. Albany: State University of New York Press, 1992. 151–83.

Hartley, Beryl. "Exploring and Communicating Knowledge of Trees in the Early Royal Society." *Notes and Records of the Royal Society of London* 64.3 (2010): 229–50.

Heward, Edmund. *Matthew Hale.* London: Robert Hale, 1972.

Himma, Kenneth Einar. "Natural Law." In *The Internet Encyclopedia of Philosophy: A Peer-Reviewed Academic Resource,* ed. Bradley Dowden and James Fieser. https://www.iep.utm.edu/natlaw/.

Hinchman, Lewis, and Sandra Hinchman. "Should Environmentalists Reject the Enlightenment?" *Review of Politics* 63 (2001): 663–92.

Hirschman, Albert O. *The Passions and the Interests: Political Arguments for Capitalism before Its Triumph.* Princeton: Princeton University Press, 1977.

Hitt, Christopher. "Ecocriticism and the Long Eighteenth Century." *College Literature* 31.3 (2004): 123–47.

Holdsworth, Sir William. *A History of English Law.* Vol. 3. 4th ed. London: Methuen, 1935.

———. *A History of English Law.* Vol. 4. 3rd ed. London: Methuen, 1945.

———. *A History of English Law.* Vol. 5. 3rd ed. London: Methuen, 1945.

———. *A History of English Law.* Vol. 6. 2nd ed. London: Methuen, 1937.

———. *A History of English Law.* Vol. 7. 2nd ed. London: Methuen, 1937.

———. *A History of English Law.* Vol. 8. 2nd ed. London: Methuen, 1937.

———. *A History of English Law.* Vol. 9. 3rd ed. London: Methuen, 1944.

———. *A History of English Law.* Vol. 10. London: Methuen, 1938.

———. *A History of English Law.* Vol. 11. London: Methuen, 1938.

———. *A History of English Law.* Vol. 12. London: Methuen, 1938.

———. *A History of English Law.* Vol. 13. London: Methuen, 1952.

Holthaus, Eric. "The Kids Suing the Government over Climate Change Are Our Best Hope Now." *Slate Magazine.* November 14, 2016. https://slate.com/technology/2016/11/the-kids-lawsuit-over-climate-change-is-our-best-hope-now.html.

Hostettler, John. *The Red Gown: The Life and Works of Sir Matthew Hale.* Chichester: Barry Rose, 2002.

Huggan, Graham. "Postcolonial Ecocriticism and the Limits of Green Romanticism." *Journal of Postcolonial Writing* 45.1 (2009): 3–14.

Huggan, Graham, and Helen Tiffin. *Postcolonial Ecocriticism: Literature, Animals, Environment.* New York: Routledge, 2010.

Hutchings, Kevin. "Ecocriticism in British Romantic Studies." *Literature Compass* 4/1 (2007): 172–202.

———. *Romantic Ecologies and Colonial Cultures in the British Atlantic World, 1770–1850.* Montreal: McGill-Queen's University Press, 2009.

Iannini, Christopher P. *Fatal Revolutions: Natural History, West Indian Slavery, and the Routes of American Literature.* Chapel Hill: University of North Carolina Press, 2012.

Irlam, Shaun. "'Wish You Were Here': Exporting England in James Grainger's *The Sugar-Cane.*" *ELH* 68.2 (2001): 377–96.

Irvine, Robert P. "Labor and Commerce in Locke and Early Eighteenth-Century English Georgic." *ELH* 76.4 (2009): 963–88.

James, Susan. "Reason, the Passions, and the Good Life." In *The Cambridge History of Seventeenth-Century Philosophy,* ed. Daniel Garber and Michael Ayers. New York: Cambridge University Press, 1998. 2:1358–96.

Johns, Alessa, ed. *Dreadful Visitations: Confronting Natural Catastrophes in the Age of Enlightenment.* New York: Routledge, 1999.

Johnson, Donald R. "The Proper Study of Husbandry: Dryden's Translation of the *Georgics.*" *Restoration: Studies in English Literary Culture, 1660–1700* 6.2 (1982): 94–104.

Jones, Tom. *Pope and Berkeley: The Language of Poetry and Philosophy.* New York: Palgrave Macmillan, 2005.

———. "Pope's *Epistle to Bathurst* and the Meaning of Finance." *SEL: Studies in English Literature, 1500–1900* 44.3 (2004): 487–504.

Jones, William Powell. *The Rhetoric of Science: A Study of Scientific Ideas and Imagery in Eighteenth-Century Poetry.* Berkeley: University of California Press, 1966.

Jonsson, Fredrik Albritton. *Enlightenment's Frontier: The Scottish Highlands and the Origins of Environmentalism.* New Haven: Yale University Press, 2013.

Jordan, Nicolle. "'Where Power Is Absolute': Royalist Politics and the Improved Landscape in a Poem by Anne Finch, Countess of Winchilsea." *The Eighteenth Century* 46.3 (2005): 255–75.

Jordan, Richard Douglas. "Thomas Traherne and the Art of Meditation." *Journal of the History of Ideas* 46.3 (1985): 381–403.

Jordanova, L. J., ed. *Languages of Nature: Critical Essays on Science and Literature.* New Brunswick, NJ: Rutgers University Press, 1986.

Judge, Rebecca P. "Restoring the Commons: Toward a New Interpretation of Locke's Theory of Property." *Land Economics* 78.3 (2002): 331–38.

Kainz, Howard P. *Natural Law: An Introduction and Re-examination.* Chicago: Open Court, 2004.

Kaul, Suvir. *Poems of Nation, Anthems of Empire: English Verse in the Long Eighteenth Century.* Charlottesville: University of Virginia Press, 2000.

Keegan, Bridget. *British Labouring-Class Nature Poetry, 1730–1837.* New York: Palgrave Macmillan, 2008.

———. "Snowstorms, Shipwrecks, and Scorching Heat: The Climates of Eighteenth-Century Laboring-Class Locodescriptive Poetry." *ISLE: Interdisciplinary Studies in Literature and Environment* 10.1 (2003): 75–96.

Keenleyside, Heather. *Animals and Other People: Literary Forms and Living Beings in the Long Eighteenth Century.* Philadelphia: University of Pennsylvania Press, 2016.

———. "Personification for the People: On James Thomson's *The Seasons.*" *ELH* 76.2 (2009): 447–72.

Kelsall, Malcolm. *The Great Good Place: The Country House and English Literature.* New York: Columbia University Press, 1993.

———. "Landscapes and Estates." In *The Cambridge Companion to Alexander Pope,* ed. Pat Rogers. Cambridge: Cambridge University Press, 2007. 161–74.

Keymer, Thomas. "Weeping Dryads, Wealden Iron, and Smart's 'Against Despair': Preromantic Ecology?" *Durham University Journal* 87.2 (1995): 269–77.

Kirkham, Robert. "The Problem of Knowledge in Environmental Thought: A Counterchallenge." In *The Ecological Community: Environmental Challenges for Philosophy, Politics and Morality,* ed. Roger S. Gottlieb. London: Routledge, 1997. 193–207.

Klausen, Jimmy Casas. "Room Enough: America, Natural Liberty, and Consent in Locke's *Second Treatise.*" *Journal of Politics* 69.3 (2007): 760–69.

Koelb, Janice Hewlett. "'This Most Beautiful and Adorn'd World': Nicolson's *Mountain Gloom and Mountain Glory* Reconsidered." *ISLE* 16 (2009): 443–68.

Kramer, Cheryce, Trea Martyn, and Michael Newton, eds. *Science as Polite Culture.* Vol. 1: *Literature and Science, 1660–1834.* London: Pickering & Chatto, 2003.

Kramnick, Isaac. *Bolingbroke and His Circle: The Politics of Nostalgia in the Age of Walpole.* Cambridge, MA: Harvard University Press, 1968.

———. *Republicanism and Bourgeois Radicalism: Political Ideology in Late Eighteenth-Century England and America.* Ithaca: Cornell University Press, 1990.

Kroeber, Karl. *Ecological Literary Criticism: Romantic Imagining and the Biology of Mind.* New York: Columbia University Press, 1994.

Lamb, Jonathan. "Bruno Latour, Michel Serres, and Fictions of Enlightenment." *The Eighteenth Century: Theory and Interpretation* 57.2 (2016): 181–95.

Landry, Donna. "Georgic Ecology." In *Robert Bloomfield: Lyric, Class, and the Romantic Canon,* ed. Simon White, John Goodridge, and Bridget Keegan. Lewisburg, PA: Bucknell University Press, 2006. 253–68.

———. "Green Languages?: Women Poets as Naturalists in 1653 and 1807." *Huntington Library Quarterly* 63.4 (2000): 467–89.

————. *The Invention of the Countryside: Hunting, Walking and Ecology in English Literature, 1671–1831.* New York: Palgrave, 2001.

Latour, Bruno. *An Inquiry into Modes of Existence: An Anthropology of the Moderns.* Trans. Catherine Porter. Cambridge, MA: Harvard University Press, 2013.

————. *Politics of Nature.* Trans. Catherine Porter. Cambridge, MA: Harvard University Press, 2004.

————. *We Have Never Been Modern.* Trans. Catherine Porter. Cambridge, MA: Harvard University Press, 1993.

Lebovics, Herman. "The Uses of America in Locke's *Second Treatise on Government.*" *Journal of the History of Ideas* 47.4 (1986): 567–81.

Lee, Daniel. "Popular Liberty, Princely Government, and the Roman Law in Hugo Grotius's De Jure Belli ac Pacis." *Journal of the History of Ideas* 72.3 (2011): 371–92.

LeMenager, Stephanie, and Stephanie Foote. "The Sustainable Humanities." *PMLA* 127.3 (2012): 572–78.

Levack, Brian A. *The Civil Lawyers in England, 1603–1641: A Political Study.* Oxford: Clarendon Press, 1973.

Levine, Joseph. *Between the Ancients and the Moderns: Baroque Culture in Restoration England.* New Haven: Yale University Press, 1999.

Light, Andrew, and Eric Katz, eds. *Environmental Pragmatism.* Environmental Philosophies Series. New York: Routledge, 1996.

Loar, Christopher F. "Georgic Assemblies: James Grainger, John Dyer, and Bruno Latour." *Philological Quarterly* 97.2 (2018): 241–61.

Locher, Fabien, and Jean-Baptiste Fressoz. "Modernity's Frail Climate: A Climate History of Environmental Reflexivity." *Critical Inquiry* 38.3 (Spring 2012): 579–98.

Love, Harold. "L'Estrange, Sir Roger (1616–1704)." In *Oxford Dictionary of National Biography,* ed. H. C. G. Matthew and Brian Harrison. Oxford: Oxford University Press, 2004. Online ed. Ed. Lawrence Goldman.

Lovejoy, Arthur O. *The Great Chain of Being: A Study of the History of an Idea.* Cambridge, MA: Harvard University Press, 1936.

Low, Anthony. *The Georgic Revolution.* Princeton: Princeton University Press, 1985.

————. "New Science and the Georgic Revolution." *English Literary Reference* 13.3 (Autumn 1983): 231–59.

Lupton, Christina, Sean Silver, and Adam Sneed. "Introduction: Latour and Eighteenth-Century Studies." *The Eighteenth Century: Theory and Interpretation* 57.2 (2016): 165–79.

Lussier, Mark. *Romantic Dynamics: The Poetics of Physicality.* New York: St. Martin's Press, 2000.

MacIntyre, Alasdair. *After Virtue: A Study in Moral Theory.* Notre Dame: University of Notre Dame Press, 1981.

————. *A Short History of Ethics: A History of Moral Philosophy from the Homeric Age to the Twentieth Century*. 1967. 2nd ed. London: Routledge, 1998.

Mack, Maynard. *Alexander Pope: A Life*. New Haven: Yale University Press, 1985.

————. *The Garden and the City: Retirement and Politics in the Later Poetry of Pope, 1731–1743*. Toronto: University of Toronto Press, 1969.

Maclean, Gerald, Donna Landry, and Joseph P. Ward, eds. *The Country and the City Revisited: England the Politics of Culture, 1550–1850*. New York: Cambridge University Press, 1999.

Macpherson, C. B. *The Political Philosophy of Possessive Individualism: Hobbes to Locke*. Oxford: Clarendon Press, 1962.

Maresca, Thomas. *Pope's Horatian Poems*. Columbus: Ohio State University Press, 1966.

Markley, Robert. *Fallen Languages: Crises of Representation in Newtonian England, 1660–1740*. Ithaca, NY: Cornell University Press, 1993.

Marris, Emma. *Rambunctious Garden: Saving Nature in a Post-Wild World*. New York: Bloomsbury, 2011.

Marsh, George Perkins. *Man and Nature; or, Physical Geography as Modified by Human Action*. New York: Scribner, 1864.

McGee, J. Sears. "Adams, Thomas (1583–1652)." In *Oxford Dictionary of National Biography*. Oxford: Oxford University Press, 2004. Online ed. Ed. Lawrence Goldman.

McKusick, James C., ed. *Green Writing: Romanticism and Ecology*. New York: St. Martin's, 2000.

————. "John Evelyn: Forestry of Imagination." *English Faculty Publications* 17 (2013): n.p. http://scholarworks.umt.edu/eng_pubs/17.

McLean, A. J. "The Common Law Life Estate and the Civil Law Usufruct: A Comparative Study," *International and Comparative Law Quarterly* 12.2 (1963): 649–67.

McRae, Andrew. *God Speed the Plough: The Representation of Agrarian England*. New York: Cambridge University Press, 1996.

Menely, Tobias. *The Animal Claim: Sensibility and the Creaturely Voice*. Chicago: Chicago University Press, 2015.

————. "Animal Signs and Ethical Significance: Expressive Creatures in the British Georgic." *Mosaic: A Journal for the Interdisciplinary Study of Literature* 39.4 (2006): 111–18.

————. "'The Present Obfuscation': Cowper's *Task* and the Time of Climate Change." *PMLA* 127.3 (2012): 477–92.

————. "Sovereign Violence and the Figure of the Animal, from *Leviathan* to *Windsor-Forest*." *Journal for Eighteenth-Century Studies* 33.4 (2010): 567–82.

Mentz, Steve. "After Sustainability." *PMLA* 127.3 (2012): 586–92.

Merchant, Carolyn. *The Death of Nature: Women, Ecology and the Scientific Revolution.* San Francisco: Harper and Row, 1983.

———. *Radical Ecology: The Search for a Livable World.* New York: Routledge, 1992.

Mikhail, Alan. "Enlightenment Anthropocene." *Eighteenth-Century Studies* 49.2 (2016): 211–31.

Miles, George B. "'Georgics' 3.209–294: 'Amor' and Civilization." *California Studies in Classical Antiquity* 8 (1975): 177–97.

———. *Virgil's Georgics: A New Interpretation.* Berkeley: University of California Press, 1980.

Milne, Anne. *"Lactilla Tends Her Fav'rite Cow": Ecocritical Readings of Animals and Women in Eighteenth-Century British Labouring-Class Women's Poetry.* Lewisburg, PA: Bucknell University Press, 2008.

Milton, J. R. "Laws of Nature." In *The Cambridge History of Seventeenth-Century Philosophy*, ed. Daniel Garber and Michael Ayers. New York: Cambridge University Press, 1998. 1:680–701.

Moore, Jason. "The End of the Road?: Agricultural Revolutions in the Capitalist World-Ecology, 1450–2010." *Journal of Agrarian Change* 10.3 (2010): 389–413.

———. "Sugar and the Expansion of the Early Modern World Economy: Commodity Frontiers, Ecological Transformation, and Industrialization." *Review* 23.3 (2000): 406–33.

Morton, Timothy. *Ecology without Nature: Rethinking Environmental Aesthetics.* Cambridge, MA: Harvard University Press, 2007.

———. "Environmentalism." In *Romanticism: An Oxford Guide*, ed. Nicholas Roe. New York: Oxford University Press, 2005.

Murphy, Mark. "The Natural Law Tradition in Ethics." In *The Stanford Encyclopedia of Philosophy*, ed. Edward N. Zalta. Summer 2019 ed. https://plato .stanford.edu/archives/sum2019/entries/natural-law-ethics.

Myers, Katherine. "Shaftesbury, Pope, and Original Sacred Nature." *Garden History* 38.1 (2010): 3–19.

Noggle, James. "Taste and Temporality in *An Epistle to Burlington*." *Studies in the Literary Imagination* 38.1 (2005): 117–35.

Nozick, Robert. *Anarchy, State, and Utopia.* New York: Basic Books, 1974.

O'Brien, Karen. "Imperial Georgic, 1660–1789." In *The Country and the City Revisited*, ed. Gerald Maclean, Donna Landry, and Joseph P. Ward. New York: Cambridge University Press, 1999. 160–79.

Onuf, Peter S., ed. *Jeffersonian Legacies.* Charlottesville: University of Virginia Press, 1993.

Orr, David W. *Dangerous Years: Climate Change, the Long Emergency, and the Way Forward.* New Haven: Yale University Press, 2016.

Parker, Blanford. *The Triumph of Augustan Poetics: English Literary Culture from Butler to Johnson.* New York: Cambridge University Press, 1998.

Passmore, John. *Man's Responsibility for Nature: Ecological Problems and Western Traditions.* London: Duckworth, 1980.

Pellicer, Juan Christian. "Celebrating Queen Anne and the Union of 1707 in Great Britain's First Georgic." *Journal for Eighteenth-Century Studies* 37.2 (2014): 217–27.

———. "The Georgic at Mid-Eighteenth Century and the Case of Dodsley's 'Agriculture.'" *Review of English Studies* 54.213 (2003): 67–93.

———. "Harleian Georgic from Tonson's Press: The Publication of John Philips's *Cyder*, 29 January 1708." *The Library* 7.2 (2006): 185–98.

———. "Introduction: The Politics of *Cyder*." In *Cyder: A Poem in Two Books*, by John Philips. 1708. Ed. John Goodridge and J. C. Pellicer. Cheltenham, England: The Cyder Press, 2001. i–xviii.

———. "Reception, Wit, and the Unity of Virgil's *Georgics*." *Symbolae Osloenses: Norwegian Journal of Greek and Latin Studies* 82 (2007): 90–115.

Perkell, Christine G. *The Poet's Truth: A Study of the Poet in Virgil's Georgics.* Berkeley: University of California Press, 1989.

Peterfreund, Stuart. "From the Forbidden to the Familiar: The Way of Natural Theology Leading Up to and Beyond the Long Eighteenth Century." *Studies in Eighteenth-Century Culture* 37 (2008): 23–39.

Philippon, Daniel J. "Sustainability and the Humanities: An Extensive Pleasure." *American Literary History* 24.1 (2012): 163–79.

Pickard, Richard. "Environmentalism and 'Best Husbandry': Cutting Down Trees in Augustan Poetry." In *Theatre of the World/Théâtre du monde*. Ed. Carol Gibson-Wood and Gordon D. Fulton. Edmonton: Academic, 1998. 103–26.

Pincus, Steven. "John Evelyn: Revolutionary." In *John Evelyn and His Milieu*, ed. Francis Harris and Michael Hunter. London: The British Library, 2003. 185–220.

Plumwood, Val. *Environmental Culture: The Ecological Crisis of Reason.* New York: Routledge, 2002.

———. *Feminism and the Mastery of Nature.* New York: Routledge, 1993.

Pocock, J. G. A. *The Ancient Constitution and the Feudal Law: A Study of English Historical Thought in the Seventeenth Century: A Reissue with a Retrospect.* Cambridge: Cambridge University Press, 1987.

———. *The Machiavellian Moment: Florentine Political Thought and the Atlantic Republican Tradition.* Princeton: Princeton University Press, 1973.

———. *Virtue, Commerce, and History: Essays on Political Thought and History, Chiefly in the Eighteenth Century.* Cambridge: Cambridge University Press, 1985.

Pollock, Sir Frederick. *The Land Laws.* 3rd ed. London: MacMillan and Co., 1896.

Pope Francis. "Laudato si." 24 May 2015. http://w2.vatican.va/content/francesco
/en/encyclicals/documents/papa-francesco_20150524_enciclica-laudato-si
.html.

Popkin, Richard. "The Religious Background of Seventeenth-Century Philosophy." In *The Cambridge History of Seventeenth-Century Philosophy*, ed. Daniel
Garber and Michael Ayers. New York: Cambridge University Press, 1998.
1:393–422.

Quehen, Hugh de. "Prideaux, Humphrey (1648–1724)." In *Oxford Dictionary
of National Biography*, ed. H. C. G. Matthew and Brian Harrison. Oxford:
Oxford University Press, 2004. 45:341–42.

Rackham, Oliver. *The History of the Countryside*. London: J. M. Dent & Sons,
1986.

Randhawa, Beccie Puneet. "The Inhospitable Muse: Locating Creole Identity
in James Grainger's *The Sugar-Cane*." *The Eighteenth Century* 49.1 (2008):
67–85.

Rawlings, Philip. "Dodd, William (1729–1777)." In *Oxford Dictionary of National Biography*, ed. H. C. G. Matthew and Brian Harrison. Oxford: Oxford
University Press, 2004. 16:400–402.

Richards, John F. *The Unending Frontier: An Environmental History of the Early
Modern World*. Berkeley: University of California Press, 2003.

Rival, Laura. "Trees, from Symbols of Life and Regeneration to Political Arte-
facts." In *The Social Life of Trees: Anthropological Perspectives on Tree Symbolism*, ed. Laura Rival. Oxford: Berg, 1998. 1–36.

Roby, Henry John. *An introduction to the study of Justinian's Digest, containing an
account of its composition and of the jurists used or referred to therein, together
with a full commentary on one title (De usufructu)*. Cambridge: The University
Press, 1884.

Roemer, John E. "A Challenge to Neo-Lockeanism." *Canadian Journal of Philosophy* 18.4 (1988): 697–710.

Rogers, Pat. "John Philips, Pope, and Political Georgic." *Modern Language
Quarterly* 66.4 (2005): 411–42.

———. *A Political Biography of Alexander Pope*. London: Pickering and Chatto,
2010.

Roxburgh, Natalie. "Rethinking Gender and Virtue through Richardson's Domestic Accounting." *Eighteenth-Century Fiction* 24.3 (2012): 403–29.

Rule, John. *Albion's People: English Society, 1714–1815*. Social and Economic History of England. New York: Longman, 1992.

———. *The Vital Century: England's Developing Economy, 1714–1815*. Social and
Economic History of England. New York: Longman, 1992.

Saunders, Beatrice. *John Evelyn and His Times*. Oxford: Pergamon Press, 1970.

Sax, Joseph L. "The Public Trust Doctrine in Natural Resource Law: Effective
Judicial Intervention." *Michigan Law Review* 68 (1970): 471–566.

Schama, Simon. *Landscape and Memory.* New York: Knopf, 1995.

Schmidgen, Wolfram. *Eighteenth-Century Fiction and the Law of Property.* New York: Cambridge University Press, 2002.

Schweiger, Tristan J. "Grainger's West Indian Planter: Property and Authority in *The Sugar-Cane.*" *Eighteenth-Century Studies* 50.4 (Summer 2017): 401–16.

Serres, Michel. "The Natural Contract." Trans. Felicia McCarren. *Critical Inquiry* 19 (1992): 1–21.

———. *The Natural Contract.* Trans. Elizabeth MacArthur and William Paulson. Ann Arbor: University of Michigan Press, 1995.

Setzer, Sharon. "'Pond'rous Engines' in 'Outraged Groves': The Environmental Argument of Anna Seward's 'Colebrook Dale.'" *European Romantic Review* 18.1 (2007): 69–82.

Shannon, Laurie. *The Accommodated Animal: Cosmopolity in Shakespearean Locales.* Chicago: University of Chicago Press, 2013.

Sharpe, Kevin, and Steven N. Zwicker, eds. *Refiguring Revolutions: Aesthetics and Politics from the English Revolution to the Romantic Revolution.* Berkeley: University of California Press, 1998.

Shields, David S. *Oracles of Empire: Poetry, Politics, and Commerce in British America, 1690–1750.* Chicago: University of Chicago Press, 1990.

Shrader-Frechette, Kristin. "Locke and Limits on Land Ownership." *Journal of the History of Ideas* 54.2 (1993): 201–19.

Silver, Sean. "Hooke, Latour, and the History of Extended Cognition." *The Eighteenth Century: Theory and Interpretation* 57.2 (2016): 197–215.

Simmons, I. G. *An Environmental History of Great Britain.* Edinburgh: Edinburgh University Press, 2001.

Sitter, John. *The Cambridge Introduction to Eighteenth-Century Poetry.* New York: Cambridge University Press, 2011.

———. "Ecological Prospects and Natural Knowledge." In *The Cambridge Introduction to Eighteenth-Century Poetry,* 198–215.

———. "Eighteenth-Century Ecological Poetry and Ecotheology." *Religion and Literature* 40.1 (2008): 1–26.

Sloan, Herbert E. "'The Earth Belongs in Usufruct to the Living.'" In *Jeffersonian Legacies,* ed. Peter S. Onuf. Charlottesville: University of Virginia Press, 1993. 281–315.

———. *Principle and Interest: Thomas Jefferson and the Problem of Debt.* New York: Oxford University Press, 1995.

Smith, Courtney Weiss. "Anne Finch's Descriptive Turn." *The Eighteenth Century: Theory and Interpretation* 57.2 (2016): 251–65.

———. *Empiricist Devotions: Science, Religion, and Poetry in Early Eighteenth-Century England.* Charlottesville: University of Virginia Press, 2016.

———. "Political Individuals and Providential Nature in Locke and Pope." *SEL: Studies in English Literature 1500–1900* 52.3 (2012): 609–29.

Smith, William. *A Dictionary of Greek and Roman Antiquities*. London: John Murray, 1875.

Smout, T. C. *Nature Contested: Environmental History in Scotland and Northern England since 1600*. Edinburgh: Edinburgh University Press, 2000.

Spacks, Patricia Meyer. *Reading Eighteenth-Century Poetry*. Malden, MA: Blackwell, 2009.

Spurr, John. "Allestree, Richard (1621/2–1681)." In *Oxford Dictionary of National Biography*. Online ed. Ed. Lawrence Goldman. Oxford: Oxford University Press, 2004.

Squadrito, Kathleen. "Locke's View of Dominion." *Environmental Ethics* 1.3 (1979): 255–62.

Stack, Frank. *Pope and Horace: Studies in Imitation*. Cambridge: Cambridge University Press, 1985.

Stein, Peter. *The Roman Law in European History*. Cambridge: Cambridge University Press, 1999.

Stewart, M.A. "The Curriculum in Britain, Ireland and the Colonies." In *The Cambridge History of Eighteenth-Century Philosophy*. Ed. Knud Haaksonssen. New York: Cambridge University Press, 2006. 1:97–120.

Stranks, C. J. *Anglican Devotion: Studies in the Spiritual Life of the Church of England between the Reformation and the Oxford Movement*. London: SCM Press, 1961.

Strauss, Leo. *Natural Right and History*. Chicago: University of Chicago Press, 1953.

Sturm, Douglas. "Property: A Relational Perspective." *Journal of Law and Religion* 4.2 (1986): 353–404.

Tennant, R. C. "Christopher Smart and *The Whole Duty of Man*." *Eighteenth-Century Studies* 13.1 (1979): 63–78.

Theis, Jeffrey S. *Writing the Forest in Early Modern England: A Sylvan Pastoral Nation*. Pittsburgh: Duquesne University Press, 2009.

Thomas, Keith. *Man and the Natural World*. London: Penguin Books, 1984.

Thompson, E. P. *The Making of the English Working Class*. New York: Vintage Books, 1966.

Thompson, Helen. *Ingenuous Subjection: Compliance and Power in the Eighteenth-Century Domestic Novel*. Philadelphia: University of Pennsylvania Press, 2005.

———. *Customs in Common: Studies in Traditional Popular Culture*. New York: The New Press, 1992.

Trachtenberg, Zev. "John Locke: 'This Habitable Earth of Ours.'" In *Engaging Nature: Environmentalism and the Political Theory Canon*, ed. Peter Cannavò and Joseph H. Lane Jr. Cambridge, MA: MIT Press, 2014.

Tully, James. *A Discourse on Property: John Locke and His Adversaries*. New York: Cambridge University Press, 1980.

Turner, James. *The Politics of Landscape: Rural Scenery and Society in English Poetry*. Cambridge, MA: Harvard University Press, 1979.

Waldron, Jeremy. *God, Locke and Equality: Christian Foundations of John Locke's Political Thought*. New York: Cambridge University Press, 2002.

———. *The Right to Private Property*. New York: Oxford University Press, 1988.

Ward, W. R. "Horneck, Anthony (1641–97)." In *Oxford Dictionary of National Biography*, ed. H. C. G. Matthew and Brian Harrison. Oxford: Oxford University Press, 2004. 28:155–56.

Warde, Paul. "The Invention of Sustainability." *Modern Intellectual History* 8.1 (2011): 153–70.

———. *The Invention of Sustainability: Nature and Destiny, c.1500–1870*. Cambridge: Cambridge University Press, 2018.

Wasserman, Earl R. *Pope's Epistle to Bathurst: A Critical Reading with an Edition of the Manuscripts*. Baltimore: Johns Hopkins University Press, 1960.

Watson, Fiona. "New Directions in Scottish History: Environmental History." *Scottish Historical Review* 82.2 (2003): 285–94.

Watson, Robert N. *Back to Nature: The Green and the Real in the Late Renaissance*. Philadelphia: University of Pennsylvania Press, 2006.

White, Lynn, Jr. "The Historical Roots of Our Ecologic Crisis." *Science* 155 (1967): 1203–7.

Wilkinson, L. P. *The Georgics of Virgil: A Critical Survey*. London: Cambridge University Press, 1969.

Willey, Basil. *The Eighteenth-Century Background: Studies in the Idea of Nature in the Thought of the Period*. New York: Columbia University Press, 1941.

Williams, Aubrey L. "A Hell for 'Ears Polite': Pope's *Epistle to Burlington*." *ELH* 51.3 (1984): 479–503.

Williams, Raymond. *The Country and the City*. New York: Oxford University Press, 1973.

Wirth, Thomas. "'So Many Things for His Profit and for His Pleasure': British and Colonial Naturalists Respond to an Enlightenment Creed, 1727–1777." *Pennsylvania Magazine of History and Biography* 131 (2007): 127–47.

Wolf, Clark. "Contemporary Property Rights, Lockean Provisos, and the Interests of Future Generations." *Ethics* 105.4 (1995): 791–818.

———. "Property Rights, Human Needs, and Environmental Protection: A Response to Brock." *Ethics and the Environment* 4.1 (1999): 107–113.

Wolloch, Nathaniel. "The Civilizing Process, Nature, and Stadial Theory." *Eighteenth-Century Studies* 44.2 (2011): 245–59.

———. *History and Nature in the Enlightenment: Praise of the Mastery of Nature in Eighteenth-Century Historical Literature*. Burlington, VT: Ashgate, 2011.

Wood, Gillen D'Arcy. "Introduction: Eco-historicism." *Journal for Early Modern Cultural Studies* 8.2 (2008): 1–7.

————. "What Is Sustainability Studies?" *American Literary History* 24.1 (2012): 1–15.

Wood, Neal. *John Locke and Agrarian Capitalism.* Berkeley: University of California Press, 1984.

Worster, Donald. *Nature's Economy: A History of Ecological Ideas.* New York: Cambridge University Press, 1977.

————. *The Wealth of Nature: Environmental History and the Ecological Imagination.* New York: Oxford University Press, 1993.

Wrightson, Keith. *Earthly Necessities: Economic Lives in Early Modern Britain.* The New Economic History of Britain. New Haven: Yale University Press, 2000.

Zarka, Yves Charles. "The Foundations of Natural Law." *British Journal for the History of Philosophy* 7.1 (1999): 15–32.

Zuckert, Michael P. *Natural Rights and the New Republicanism.* Princeton: Princeton University Press, 1994.

INDEX

CPSIA information can be obtained
at www.ICGtesting.com
Printed in the USA
LVHW011655091221
705747LV00007B/1255